→ Fall prevent'n idea - d[...]
 'work more' on RCAs o
 (incr. emphasis)

2's input of a fall

→ Shell execs manage from a   boss
 idea - still deliver, despite [...] - system must put in place

→ 3 Qus idea great 99-100
 And engaging w/ 'workers' - big them up vs.
 arrogant, ivory tower process designers
 - like the "Depot listening process"

PRE-ACCIDENT INVESTIGATIONS

WORKPLACE FATALITIES
Failure To Predict

→ Hazards not fixed - can't manage them "in an
 office." / a "rigid process"
→ Fundamental shift - "What do I do, that makes
 your job harder/easier?" snr/office based
            Todd Conklin, PhD   qus to a 'worker'

Data?   Santa Fe, New Mexico

— Fatality #s over 25yrs + lower cons eq incidents
— Finding for TRIPODS - workers failed to: 10 hazard.
                                    Follow process

— When an incident of high severity - how
 often/deeply does SteM ask - what do
 the people involved need?

Copyright © 2017 Todd E. Conklin

All rights reserved.

ISBN-10: 1546979454
ISBN-13: 97815469654

DEDICATION

To all the workers who have given their lives to simply get a job done. Honoring you by learning and improving is the most important thing we can do.

CONTENTS

Foreword	Sidney Dekker	i
Prelude: The Flip of a Coin		8
Failure to Prevent...		10
Fatalities and Serious Incidents		25
Fatalities are Not Normal		31
Fatalities are Anomalies		36
How Fatalities Differ from Other Events		42
There is a Bias to Prevention		46
Our Discussion So Far		69
The Great Shift in Thinking		75
Risk Defined and Discussed		99
The Space Where Worker Manage Risk		109
What we Know Now		119
The Mission: Stop Killing Workers		125
Postlude: The Power of Restoration		128

Foreword

Many organizations tell us that work has never been as safe as it is today. They will show the lowest injury figures ever, and the rosiest incident counts in years. They want to be proud of these accomplishments, and perhaps they should be. But behind these results hides complexity and contradiction—a messiness that Todd Conklin takes us into with this book. For one, it is pretty obvious by now that trying to lower our incident and injury rates leaves the risk of process safety disasters and fatalities pretty much unaffected. Getting better at managing injuries and incidents doesn't help us prevent fatalities and accidents—we've known that for a long time (Salminen, Saari, Saarela, & Rasanen, 1992). The number of fatalities in, say, construction, or the energy industry, has remained relatively stable over the past decades (Amalberti, 2013; National-Safety-Council, 2004), even when many organizations proudly report entire years (or more) without injury. Lowering the injury or non-serious incident rate can actually put an organization at greater risk of accidents and fatalities. In shipping, for example, injury counts were halved over a recent decade, but the number of shipping accidents tripled (Storkersen, Antonsen, & Kongsvik, 2016). In construction, most workers lost their lives precisely in the years with the lowest injury counts (Saloniemi & Oksanen, 1998). And in aviation, airlines with the fewest incidents have the highest passenger mortality risk (Barnett & Wang, 2000).

What lies behind these fatalities? Do they really happen because some people don't wear their personal protective equipment; that some don't wear gloves when rules say they should? We probably all know this example:

> For years BP had touted its safety record, pointing to a steep decline in the number of slips, falls, and vehicle accidents that generate days away from work, a statistic that is closely followed by both the industry and its regulators. BP had established a dizzying array of rules

that burnished this record, including prohibitions on driving while speaking on a cellphone, walking down a staircase without holding a handrail, and carrying a cup of coffee around without a lid. Bonuses for BP executives included a component tied to these personal-injury metrics. BP cut its injury rate dramatically after the Amoco merger [the previous owner of the Texas City refinery]. But BP's personal-safety achievements masked failures in assuring process safety. In the energy business, process safety generally comes down to a single issue: keeping hydrocarbons contained inside a steel pipe or tank. Disasters don't happen because someone drops a pipe on his foot or bumps his head. They result from flawed ways of doing business that permit risks to accumulate (Elkind, Whitford, & Burke, 2011, p. 7).

Large-scale, serious, life-changing injuries and fatalities have remained stubbornly constant in these and other industries for the last twenty years or more—even if these numbers are historically low by comparison. In some cases, they have even been on the increase. Perhaps we were always amazingly naïve to think they would budge under interventions that target workers and their behavior. Of course, some who work in safety keep claiming just that, despite the clear data: More people-centered rules do not create more safety for systems or processes, and sometimes not even for people (Amalberti, 2013). In mining accidents in New Zealand and the US, for example, workers would have been warned of sanction if they didn't don their yellow vests on site. But then 29 of them died in a mine collapse in each country. In another example, workers had to strictly follow driving and walking regulations on a chemical plant site, but then four of them died in a toxic gas release in a building on that same site—two of them brothers (Hlavaty, Hassan, & Norris, 2014). And workers for a copper mine were taking part in a compulsory behavioral safety course in an underground training facility. Then the roof of the tunnel in which they were gathered collapsed. It killed 28 miners and injured 10—while they were being trained in behavioral safety (Santhebennur, 2013). Such safety interventions are intended to reduce unsafe acts, to weed out unwanted worker behaviors, to limit higher-frequency/lower-consequence incidents, or to prevent minor injuries—and they have not had an influence on fatalities or major incidents. The only influence is that they sometimes increase the chance or prevalence of the really big, bad stuff.

Paradoxically, dropping our obsession with making sure everything goes right—and finally giving up on targeting the worker's behavior to make it so—is probably a very good place to start. Investigating and understanding daily success can actually reveal where the next potential adverse outcomes might come from. And it can do that much better than investigating the highly infrequent failure. The reason for that seems to be this. An organization that has already achieved a pretty good safety record evidently has got its known sources of risk under acceptable control. But the types of accidents that might still happen in these organizations, as Amalberti (2001) argued, are no longer preceded by the sorts of incidents that get formally flagged or reported. Instead, accidents are preceded by normal, daily, successful work. This will likely include the so-called 'workarounds' and daily frustrations, the improvisations and adaptations, the shortcuts, as well as the sometimes unworkable or unfindable tools, vexing technologies and the occasionally unreliable results or readings from various instruments and measurements. These things are typically not reported: they are just all part of the daily creation of safety despite an imperfect, non-deterministic world. It's all in the game. People have learned to live with it, work around it, and get things done. Leaders need to learn about these things, because they tend to be the conditions that allow their organization to drift into failure, and that show up in the fatalities that still do happen (Dekker, 2011). We can't obviously learn about these conditions if we threaten with sanctions when not all the rules are followed precisely. That will shut people up for as long as we are there: they'll temporarily halt the workarounds and little innovations and improvisations, which normally get stuff done. To learn how work is actually done—as opposed to how we think it is done—we need an open mind, and a big heart. And we need to take our time, and use our ears more than our mouths.

And how should we respond when, very occasionally, things do go terribly wrong? The typical response that organizations have is a pretty retributive one. They will ask questions such as:

- What rule was broken?
- Who is responsible?
- How bad is the violation (honest mistake, at-risk or reckless behavior) and so what should the consequences be?

Such a response is organized around shades of retribution. It focuses on the single 'offender,' or perhaps a small team, asks what they have done and what they deserve. But, and Todd Conklin has said this often too, learning and punishing are mutually exclusive. Meeting hurt with even more hurt also doesn't make us better people, or better leaders. There has to be a better way. A restorative approach to the aftermath of an accident or fatality asks very different questions:

- Who is hurt?
- What are their needs?
- Whose obligation is it to meet those needs?

Such an approach is not only more 'just,' but also more inclusive. A variety of people can get affected by an incident: not just the first victims (e.g. the worker) but also colleagues, supervisors, bystanders, the organization, the surrounding community—they too may somehow have been affected by what happened. If, as fellow human beings, and especially as leaders, we realize that hurt creates needs, and that needs create obligations, then we can take our responses and conversations in a very different direction. Restorative approaches succeed by systematically considering those needs, and working out collaboratively whose obligation it is to meet them. When fatalities do happen, and an organization is reeling from the shock and grief, the last thing we should do is to revert to blame. Let's not hunt for scapegoats on whom we can pin some imaginary 'cause,' since we'll have it wrong anyway or we will grossly oversimplify the causal web leading up to the bad outcome. What we need to do instead is ask different questions, invite different conversations. In the wake of terrible loss and fatality, we need to restore what has just been broken and lost: the trust and the sense of community; the realization that we are all in this together—workers, managers, leaders; the recognition that only if we come together can we get better at avoiding these terrible things.

Professor Sidney Dekker

Director, Safety Science Innovation Lab, Griffith University, Brisbane

References

Amalberti, R. (2001). The paradoxes of almost totally safe transportation systems. *Safety Science, 37*(2-3), 109-126.

Amalberti, R. (2013). *Navigating safety: Necessary compromises and trade-offs -- theory and practice*. Heidelberg: Springer.

Barnett, A., & Wang, A. (2000). Passenger mortality risk estimates provide perspectives about flight safety. *Flight Safety Digest, 19*(4), 1-12.

Dekker, S. W. A. (2011). *Drift into failure: From hunting broken components to understanding complex systems*. Farnham, UK: Ashgate Publishing Co.

Elkind, P., Whitford, D., & Burke, D. (2011, 24 January). BP: 'An accident waiting to happen'. *Fortune, 85*, 1-14.

Hlavaty, C., Hassan, A., & Norris, M. (2014). Investigation begins into 4 workers deaths at La Porte plant. *Houston Chronicle, Sunday*(November 16), 1-4.

National-Safety-Council. (2004). *Injury facts 2004 edition*. Retrieved from Itasca, IL:

Salminen, S., Saari, J., Saarela, K. L., & Rasanen, T. (1992). Fatal and non-fatal occupational incidents: Identical versus differential causation. *Safety Science, 15*, 109-118.

Saloniemi, A., & Oksanen, H. (1998). Accidents and fatal accidents: Some paradoxes. *Safety Science, 29*, 59-66.

Santhebennur, M. (2013). Picking up the pieces: Indonesian mine collapse. *Australian Mining, 25,* 4-5.

Storkersen, K., Antonsen, S., & Kongsvik, T. (2016). One size fits all? Safety management regulation of ship accidents and personal injuries. *Journal of Risk Research*. doi:http://dx.doi.org/10.1080/13669877.2016.1147487

Prelude: The Flip of a Coin

Before you begin to read this book, go get a coin. This book talks a lot about coin flipping, so having and flipping your coin will set the tone and give some context to how we are going to talk about serious outlier events.

Any coin will work. The only requirement for the coin is that you are able to flip it. I want you to participate in a small experiment before we start to have a conversation about fatality and serious events, catastrophic failure of any type. Do this micro-experiment alone, do this experiment with friends. Just do me this one favor, flip your coin enough times seriously create an environment in which you can learn.

Flip your coin several times and keep track of whether the coin lands heads up or tails up. Be patient, you will need to flip this coin probably 20 or 30 times in order to begin to get an idea of how probabilities work. Stay strong and keep flipping until you begin to notice a pattern of outcomes.

Eventually, all of this coin flipping will begin to produce a predictable outcome. What seems so driven by chance, luck if you will even though I am not very fond of the word, will begin to create collective predictability. It won't happen right away, but soon enough you will see a result of all of your efforts. You will notice that the number of times the coin comes up heads begins to match the number of time the coin comes up tails. Soon, it will become very clear that the odds for flipping a coin are consistently and predictably 50/50.

I can be certain of the outcome of a coin flip over time. If I flip the coin hundreds of times I know the odds between heads and tails will be 50/50. Half of the flips will be heads and half of the flips will be tails. I know this is true. Now, because of your micro-experiment, you also know this is true.

I promise this is true. Set your watch to it, you can depend on this outcome every time. If you flip a coin enough times you will always get this very dependable and predictable outcome.

Except there is this one little problem...

This little problem is best shared with you by asking this question, "can you really depend on this outcome every time? Can you bet all the money in your bank account on this outcome every time? Would you bet your life on this outcome every time?"

The problem is that I know the outcome of a coin flip *over time*. You give me enough flips of the coin and I can absolutely know and predict the future with certainty. I know what happens when I flip a coin over time - I just can't know the outcome of the *next* flip of the coin. I can't predict the outcome the next time.

Discussing a coin flip is an easily shown and an easily shared experience for the basis for this book. I know what my operations can do overtime. I just can't know what my operation will do next time. I can't know what will happen the next time my organization does high-risk work.

Welcome to our journey into knowing that the one thing we can't manage is uncertain outcomes. Not the long-term uncertainty that we have managed and will manage over time, but the uncertainty of the next operation. Since we cannot manage uncertainty – we must manage our reaction and preparation for uncertain outcomes.

Enjoy this conversation.

Part One

Failure To Prevent Vs. Failure to Control?

"Why a book about fatalities and serious incidents?" my colleague asked.

To be honest, I don't really know. I certainly did not choose to spend my time thinking about workers who die while doing work.

I am hesitant to write this book and unless someone has had a fatality happen in their workplace, I am not sure why anyone would want to read this book. Who would want to even think about such an awful outcome? Another reason I am hesitant is that I am also certain the point of view I have on this problem in not going to be very popular.

My answer is best given in a couple of global and somewhat disconnected reasons. Work is complex! No explanation for how a worker does his or her work is ever simple, and there is a certain amount of context always begging for further explanation. Nothing about fatality and serious incident work is simple either. There are no easy ways to answer the question, "How could the worst possible outcome happen in my organization?" Yet there are any number of companies starting to appear in the safety and reliability community who seem to be selling some potentially dangerous and overly simplified answers. I am fairly certain you cannot buy answers as to why people die while doing work. As Admiral Hyman Rickover (the father of the US Nuclear Navy and an early thinker in safety and reliability) is credited to have said, "There are no easy answers to hard problems". This problem is not a simple problem. Unfortunately, the solution to this problem won't be simple either.

A discussion, a conversation with a purpose, about fatalities and serious events seems more than fitting for those of us faced with these problems in today's workplace. This is only the beginning of the discussion about fatalities and serious events, their origins and their remedies; at least this is an early discussion for me. It is my hope the discussion does not serve to provide simple and shallow answers to this complicated problem. It is also my hope this small book will help frame the thinking and, therefore, the actions that your organization can immediately make around the important and never-ending quest to stop killing people and having catastrophic outcomes.

I know one thing for sure: Doing what we have been doing – harder and better – will not solve the problem of fatalities at work. In fact, I would propose the idea of doing more of the same-old-same-responses might just be the problem. Continuing to think about and act on fatalities and serious events using our traditional thinking almost guarantees more people will die or get seriously injured. If we keep doing what we have always done, we're always going to get what we have right now. What we have right now is not working.

Reason one: There is Great Demand for this Topic.

This topic, fatalities and serious injury prevention, is beginning to get a lot of exposure and attention – which is encouraging. People have become interested in not killing or seriously injuring their workers. That is a great thing! We are starting to learn that doing the same things we have always done does not take us to any new operational results. We are also beginning to understand our current safety systems are necessary and have served us fairly well for industrial safety issues, but are not sufficient in preventing catastrophic outcomes. We know that notably safe companies and organizations, places with amazing safety programs and near-zero reportable numbers, still kill people. We have examples of excellent safety programs; yet, these programs continue to have fatalities at an unfortunately stable rate. Thus, we are being forced to think and manage our operations and systems differently in order to get new and better outcomes. We must stop the deaths of our workers. There is great demand for the answer to the question of how can we stop the deaths and injuries, a strong market pull, and yet there is so little information available to help move us all towards new thinking.

Here is the problem with having great demand for new thinking and small supply of new ideas. As the demand increases (we don't want to kill people) the "pull" on the supply (tell me what I should be doing differently) becomes greater and greater. When there is not a thoughtful and meaningful supply available (no ways to think about managing systems so the systems have the ability to consistently fail and recover), an artificial supply starts to appear on the market. his artificial supply scares me a bit. It seems in my observation the new talk about prevention of fatalities and serious incidents is just a more aggressive dose of the same stuff we have always done – do it better and do it harder and track the crap out of it. I am fairly certain doing more of the same things will not give us a different outcome. You will hear me make this case over and over.

This artificial supply (prevent bad things from happening - harder) seems wrong, really wrong. All this false information being represented as new thinking makes me worried. Good people in good organizations are going to try these old ideas-amplified and feel like their organization is going to be safer. The organization will not be safer or more stable. The organization will not prevent fatalities and serious injuries any better than before; however, the organization will think they are actively preventing deaths and catastrophic events. Worse still, organizations that do fail will spend lots of money and resources just to be told they should have been better at preventing the fatality or serious event from happening.

All of this artificial supply nonsense does not add new ideas and thoughts. We tend to want to over-simplify the problem of such horrific events in order to reinforce the same old answers. Clearly, if a person is involved in a serious event they must have done something really wrong; they missed whatever indicators were present and allowed this bad thing to happen. This person failed to prevent the event from happening. Therefore the person is the problem – the person is negligent. Right? Wrong! We continually fall in this trap and begin to believe we must reinforce and improve our event prevention thinking – and we have to do it quickly! We have meetings and we tell our workers and managers to prevent accidents more efficiently and start doing it now!

It is simply wrong to think some consultant could sell you a better way to prevent serious incidents and fatalities. I am even more worried about the other companies who are out selling quick answers that claim to make your organization better at prevention. I want everyone to be successful and I want everyone to make a living. I am certain, however, doing more prevention will not stop your workers from being seriously injured or killed.

Don't buy what the world is trying to sell you about fatalities and serious incidents; run away as fast as you can. Instead, know you will have to give up some of your current thinking in order to make room for the new thinking. Don't fix the prevention work you do – it really prevents events. Instead, know you will have to fix the ability your organization has to fail safely every time.

Which leads directly to reason two: I started seeing some interesting correlations.

I recently spent about 60 days looking at 11 fatal accidents all around the globe. I will tell you more about this journey in this book. I often get asked to go to organizations that have had a fatality or serious incident and work with the leadership team to better position the organization to learn and improve. I think of these visits as a type of restorative investigation, not in the event, but in the way the organization positions themselves to move forward. These visits are an unusual and unique opportunity to observe organizations that are in the process of recovering from operational trouble. I don't usually re-adjudicate the event or reinvestigate the accident, which would be hard to do and not super valuable.

I do spend time helping these organizations restore their ability to do high-risk work safer than the way they we were doing high-risk work before the accident. I spend hours talking with leadership and helping them rebuild their confidence in the organizations leadership ability to manage a safe and reliable operation. After all, if the leaders don't believe they will not kill people, no one in the organization will believe they have the ability to do high-risk work safely and reliably. Leaders must learn and believe they have a role in operational safety and in operational restoration. It is a very important job. I think of all the things I do in my work, this type of restorative work is most important thing I do.

I also know I am being given special access to research data that would be hard to find or recreate. I know I have a responsibility as both a scholar and practitioner to use this data in a way that challenges our thinking and pushes us as a community to move forward. Being given access to these companies means I have a responsibility to the company and to the "new view" community. I work as hard as I can to take this information, collect, correlate, and learn as much as I can.

During 2016 I was asked to work with 11 fatalities in 60 days. Doing so many of these fatality and serious incident visits in a relatively short amount of time created an opportunity I would have never have imagined could exist. I know for a fact if I went out to find this collection of data, I would not find this many different fatality events. Nor would I ever be given access. I was in a data-rich environment at precisely the perfect time to get to study a series of non-connected, but stunningly similar outcome events. I started to draw some correlations between these events. All of these events were different and yet a strong theme started to appear around all of these individual fatality events.

Believe it or not, I began to notice prevention efforts were getting in the way of recovery efforts. I was amazed. I would have never guessed this outcome would become so crystal clear; these organizations were so good at prevention they had created a false sense of security.

Here is what became crystal clear during this observation: Fatalities are preventable but not predictable. We confuse prediction and prevention and then we tend to bias our thinking, planning, and actions towards prevention. We truly believe we can prevent every fatality and yet we cannot predict every fatality or catastrophic failure.

Who would ever think the prevention efforts that failed to prevent these horrible outcomes for these companies was also potentially one of the causes of the events? Could it be an organization could get so confident in their ability to prevent fatalities they would no longer think it necessary to actually safeguard against a catastrophic event? In a way this is a remarkably counter-intuitive thought; the thought that prevention thinking erodes safeguard thinking.

I think the idea that we can predict and prevent all fatalities and serious incidents is causing us to think about the problem in a wrong way. We are falling victim of bad problem definition practices. In a way we are getting intellectually lazy – but we do not have bad solution and corrective action identification thinking. That's interesting! Our organizations are quite good and generating answers but generally bad at generating questions.

Any time I get an answer I don't understand I get curious.

Any time I am not getting the answer to a question I think should be a fairly easy question to answer, I try to remember to not blame the answer. Answers can't be bad. The answer to a question is just whatever the answer is; the answer is almost never the problem. You can't really blame and respond emotionally to the question's answer. If the answer is not the problem, then we must look at the question. Bad answers come from bad questions. That started me thinking maybe the problem was the question.

Are we asking the right question about fatalities and serious incidents?

Could it be all the work we do to prevent accidents weakens our ability to respond to accident outcomes, fatalities and serious events? Could it be over time our investigation processes and legal liability processes have caused us to limit our questions almost exclusively to only learn how we "failed to prevent a serious event" as opposed to how we "failed to control the serious outcome?" Could it be that we are confusing our prevention efforts with prediction efforts? Instead of wondering how the horrible outcome was possible, we have drifted to mostly asking how we failed to prevent the horrible outcome. We don't know how events happen. We only know what we failed to do to stop the event.

We are only getting partial information about fatalities and serious events in to our leadership offices and boardrooms. We have completely discounted the story of what happened in place of loudly telling the story of what didn't happen. Does all our emphasis on prevention set us up to be horribly surprised by the failure? Is the reason we can manage ankle sprains effectively and still kill workers because our entire thinking is around how we should prevent events from happening?

Here's what we know. We can't predict everything. We can't prevent everything. We can only hope to create systems that help us minimize the consequence. We must control the outcome. You don't manage the uncertain…you manage the certain. We don't manage accidents because accidents are uncertain. We can only manage the capacity we have in our organizations to have accidents that fail gracefully. Capacity to recover is certain, and this capacity is real.

This is our discussion. This is our conversation with a purpose. This is what we are going to explore together. Be ready! These comments will not play well in some conference rooms…we may make some people uncomfortable.

A Train, A Movie, A Crew, A Death

To begin our discussion, here is a case study of a fatal accident that happened in 2014 in the United States. This event is horrible. All fatalities are horrible. This case study may help to illustrate the conflicting ways we understand catastrophic failure after it happens. Read this case study to not only understand how the failure happened, but also to look at the way we want to, retrospectively, make a determination this accident is a failure of judgment, a failure of planning, a failure of process, and ultimately a failure to prevent the horrible outcome.

The belief that somebody did something wrong is powerful; the power of feeling certain beyond additional debate or analysis that the organization failed to prevent this outcome is an intellectually complete cause – an idea that does not require further thought or questioning. This same belief that says the horrible outcome is so complete in its cause that it does not require further discovery is limiting to the way we think and respond to accidents. It is hard to not think this event is preventable and should not have happened. This event was not prevented and did happen. Read on and see if you become interested in the obvious lack of prevention for this event. If you do, you may just be missing the much richer learning in this event.

Case Study

Sarah Jones, Cinematographer, Georgia, USA

During the production of The Midnight Rider, a biographical film about the life of rock star Gregg Allman, a 27 year old cameraperson was struck while standing on a railroad trestle bridge by an on-coming CSX freight train traveling at a speed of 60 miles per hour. Sarah Jones was a member of IATSE Local 600, working as a cinematographer at many levels in the entertainment business during her career. She was a cinematographer intern on the set of The Army Wives, as assistant cameraperson on The Vampire Diaries, and was a cinematographer on the movie The Republic of Pete. Sarah Jones was killed instantly.

It should be hard for a civilian, a member of a movie crew, to be hit and killed by a freight train. Movie crews hardly ever die in train accidents because when a film is going to do work on a railroad trestle, there are many, many controls that are normally in place before the shooting begins. In this case those controls were not insufficient; those controls that normally keep people from getting hit by a train were completely absent. There were no control or safeguards for this hazard at this worksite. How could this have happened?

The production and shooting for this scene was done in a guerilla cinema style. The productions director, Randall Miller, and his production company, Unclaimed Freight was known for their ability to create cinematic art without the normal conventions of Hollywood. Miller often used small production crews that were lean and agile to get in and out of shots that normally would take more traditional crews weeks to secure permission and photograph. There was little need to get permits or permission because the production had the ability to get in and out before anyone even knew a movie was being made. As crazy as this sounds in retrospect, it is actually done surprisingly often in the movie industry with smaller and smaller budgets and higher and higher pressure to make unique and exciting film.

Miller attracted movie workers who liked the adrenalin and action this style of movie making offered. There were no support systems on his sets. No medics or catering trucks. No fancy trailers or site security. Miller's crews could get in, shoot the scene, and get out before anybody knew what had happened. It was fun. It was a bit naughty. It was exciting, working in fear of getting caught. The problem was the fear of getting caught is different from the fear of getting killed.

Miller also attracted crews that could handle this type of fast and adaptive work. Miller needed crewmembers who were talented, able to problem-solve quickly, get the scenes and set up the film would need. There was not a lot of planning and there were no procedures and practices. The production's success depended on smart and creative crewmembers working in real time to create the movie in an artistic way. The crews were not concerned by the absence of normal conventions within the film making industry. With great sacrifice came quick advancement and almost stunning artistic reward. It was not hard for Miller to recruit his crews. His workers were not only young and talented, but also cheap to hire and energetic. The crews would work longer and more difficult hours with less support and rest then a traditional crew would require. Most importantly, the crews thrived and flourished in this environment.

Because these crews were on tight schedules and budgets and because shooting demanded light and timing, the crews ran on a high level of fatigue. Everyone worked long hours and wore those long hours like a badge of honor. In a real way, this type of shooting is more in line with the current media landscape of small, low-budget films that can go directly to the Internet.

On February 20th in Doctortown, Georgia, USA a set was quickly constructed on a railway trestle over the Altamaha River in rural Wayne County, Georgia. The shooting was of a dream sequence and involved a rather elaborate antique metal-framed bed set on the train tracks of a bridge over a canyon. The shot would be dream-like and unique. Because the shot was over a river and on a railroad trestle bridge, the fantasy aspect of this scene would be impressive.

The crew knew there would be "live trains" on this live track. The trains would not stop for the film crew; therefore, the film crew would have to stop shooting to allow the train to pass over the trestle bridge without getting caught or causing harm. Permission from CSX Railroad to use this site for this shooting of the scene was attempted, but permission was not granted by CSX Railroad. The railroad was not interested in allowing this type of activity to happen on their tracks. Permission would, most likely, never have been given to a low-budget biopic. This did not stop the crew from shooting the scene.

The film crew knew there could be a couple trains that would pass through during the shooting sequence. Two trains did come through, and the crew was able to move the actors and equipment out of the way to allow the trains to pass without harm. The crew was not aware there would be a third train or perhaps a forth or even a fifth train that day on this set of tracks. Because the crew had not secured access or permission, no CSX personnel were present to alert the crew of the "next" hazard.

When the third train was spotted the crew had about 60 seconds to move the equipment, sets, actors, and themselves out of harms way. The only escape route from the oncoming train was a metal gangplank running parallel along the track at about the midway point on the trestle. During this third iteration of the removal, the bedframe became entangled in the tracks and was more difficult to move then before. This new difficulty delayed the entire crew from getting out of the way of the speeding freight train. Not all the workers of this film crew were able to get to the safety of the gangplank.

One worker, Sarah Jones, was unable to get to the escape route and was killed. Several other workers were injured in this accident. All workers were forever impacted by this accident. There were many victims that day. Everyone was emotionally destroyed. This event had impact well beyond the actual accident. This event changed the movie making industry.

They call this type of quick in and out moving making, "scene stealing" in the movie making profession. Because of the nature of this type of filming, there were no safety personnel from either the film crew or the railroad present during the performance of this work. It was impossible to completely remove the scenery, cameras, equipment, and the crewmembers out of the danger of the third passing train without warning. Also, there was no evidence of any safety pre-planning, a safety meeting of any type, or any safety presence on the set. It should be noted none of these safety formalities were ever used for any of the production events this crew had successfully completed in the past.

The strangest thing about safe and reliable systems is that if the organization does not work to promote both prevention and recoverability, the workforce doesn't know enough to miss these two vital safety functions. This workforce was so young and inexperienced they did not know what they didn't know about safety on a film set, and their hard work, dedication to their art, and inexperience could not compensate. Because this work was always being done on the margins of safety and reliability, when those two functions are missing it does not stick out as abnormal. Workers will always normalize to the operational culture that exists, not the operational culture we desire to exist.

This event was not normal. Almost nothing about this event was like the shooting of a movie anyplace else in the world. Movies are rigid and structured events that have much coordination of many functions. That type of formality of production is important, but it is also expensive. In order to make the movie faster (and in this case bring a different aesthetic to the production) many of the normal formalities and controls that are used to ensure the safe and successful production of movie work where not in place.

However, a young and vibrant crew, a crew who actively sought this type of work, could account for many deficiencies in planning and protection. In fact, this crew had been successful up to the point where the train hit and killed one worker. They were lean and adaptive as a production advantage up and until they were too lean and adaptive to be sufficiently safe. There are many signs this work was overly risky. Those signs were not seen as dangerous because this job was simply more risky then a normal job. Risky work always has a lower-risk threshold and acceptance level. Risky work is risky.

This would be an easy event to over-simplify by investigation and analysis. A film crew should never work on a railroad trestle. If the film crew had to break the rule- "no crews working on an active track"- and actually work on a railroad trestle, the trestle should be inactive, permission should be obtained, and increased rigor and discipline around operation safety should be a minimum standard to do this work.

You could make a strong case for stop work authority here. Any one of the crewmembers should have recognized this danger and stopped this job before a member to their team was killed. It is a good bet the crew had the authority, and even if the authority had not been formally given, a crewmember could have stopped this accident before anything bad happened. It is worth noting the star of this production was reported to have said he wanted to stop this shooting; however, he was assured this low level of safety was simply "how these guys make their movies" and was encouraged to continue working in spite of his uneasy feelings toward the safety and reliability for this shoot.

There is much blame to be passed around in this fatality. Much blame was passed around, and this event ended in the court system as a negligent homicide. However, the need to talk about who did or did not do the right things for this movie production does not help us understand and learn what actually happened in this case. This fatality is more then a train hitting a worker – this fatality is a failure in the successful ability to do work. The cameraperson did not fail the production company; the production company failed the cameraperson.

Lastly, in retrospect it is easy to see where all of these mistakes and direct examples of poor reliability and safety judgment had aligned in such a way as to make this outcome seem inevitable. This event, the train hitting the crew working on the trestle, was inevitable. The problem is...in retrospect we know the train is coming with lots more time than 60 seconds of warning. A retrospective view allows us to start at the "surprise" train arrival and work backwards toward safe and reliable production. The crew did not have the benefit of retrospect and was surprised to death, literally to death, by this unknown outcome.

Being "right" in retrospect means nothing. Of course you are right; you know how this event is going to end. You have no reason to be anything but right.

The biggest problem the investigators and interested public faces in reading this case study is the belief that somehow this crew failed to predict and prevent this bad outcome from happening. That somehow and for some reason this risk was acceptable – and the crew consciously accepted this risk. Had they been better at prevention, this event would not have happened. Had they had almost any safety presence – a spotter to watch for trains – this event would not have happened.

All of those things are true and yet none of those things happened. Let me repeat this statement. All those things are true and yet none of those things happened. The question is not why didn't they prevent this event? The bigger question is how did they not have a way for the train to pass without killing members of the crew?

This case study is not about lack of prevention. Mostly what this crew did was keep bad things from happening. In fact, this crew was so good at not letting bad outcomes happen they were able to repeatedly move in and out of a hazardous environment quite safely and comfortably. If we focus on how this crew failed to prevent this accident, we are simply solving the last accident that happened, the accident no one wanted to happen. In reality, we are trying desperately to change history.

This case study is an example of an event without controls, high-risk work without the ability to fail safely. That is only part of this failure, however. This event helps to illustrate the need to move beyond a simple failure to prevent this from taking place. When planning failed, and planning fails a lot in highly adaptive work, the safeguards needed to ensure reduced consequence were not present.

Chapter 2

Fatalities and Serious Incidents

"I don't have confidence my company won't kill another person."

I was gathering up my stuff, getting ready to leave a meeting room in a fancy corporate headquarters and quickly rush to the airport to catch a plane. Standing next to the conference room table after a long and difficult presentation about a recent fatality that had happened at this organization, I was thinking about all I had just heard. I was sad and could not help but think about how suddenly life can change. How what we assume to be a dependable and stable system can change and have fatal consequence. Here one day and gone the next day. Life seemed short and fragile.

Sadly, I go to a lot of these types of meetings. I am asked to come and be a part of the thinking that goes on after a bad thing happens. Fatality investigation meetings are awful. You see photos and drawings of terrible things, blood and clothing, lists of all the things that were supposed to have happened, but didn't. You learn about the worker's family. You learn about how hurt and damaged the other workers in the organization currently feel. You realize a catastrophic event has many victims at many levels of an organization. People cry, grown people cry. It is hard to not be filled with empathy and sympathy for the organization. Senior Managers, leaders, and workers don't know how to make the story they are telling have a better ending. The problem is at this moment in time, no one has the power to rewrite the ending of this event. Fate has sealed the outcome and the best the organization can do is to react to the realities that are set forth in front of them.

There is a need to somehow try to rewrite the ending of bad events. We, as human beings, want to know quickly why some bad outcome happened. We need this information because not knowing is uncomfortable and frightening. The fear and sadness these people are feeling is real. The uncertainty in retrospect is powerful. The uncertainty for the future is scary.

During this meeting, I had just been taken through a step-by-step detailed description of a terrible accident. A worker had died…a person with a family and a future, had been killed while doing his job. Brave and smart people had let the organization and its people down in the worst possible way. Lots of things could have been different. Systems that were normally reliable somehow were not reliable. The people that would have known better were off duty or in another part of the plant. It seems like everything that could have possibly failed, failed.

These meetings are just awful. There is not much I can do but try as hard as I possibly can try to help restore the organization to a point beyond where the organization was before the accident. I use words like restore, learn, improve, move forward, and becoming a better organization because of this bad outcome. We don't really have the ability to fix or erase the bad thing that has happened. We really must use language that points the organization forward towards a better work environment.

This one specific and particular meeting was really sad and thought-provoking. This was the second fatal event in as many months for this organization. The room was filled with a feeling of severe remorse. Something this bad happening once was bad, but oddly understandable. Something this bad happening again was inexcusable and unimaginable. How could the company be this bad at safety? How come the workers didn't learn from the last fatality in order to prevent the next fatality? Why did this happen twice? What does this mean?

Those words were the very words I had used in this meeting. I gave my best effort to take this bad outcome and use this event to challenge this organization to become better than they ever have been as a company performing operations before. The organization must be restored to a place where the organization can perform work better than they did before the accident.

I was thinking while I was packing my gear I had done a good job. I don't think I was feeling "smug." I just thought I had moved this organization's conversation to a better and more effective place, a place where they could shift their mental gearbox from a retrospective reverse to a more restorative forward. I thought I had done a good job on moving the team's thinking towards a road to recovery and restoration and that change seemed true… until the most senior leader at this company took me aside and said, "I don't have confidence my company won't kill another person."

"I don't have confidence my company won't kill another person," he repeated.

My heart broke. I could not leave this office with the boss feeling a complete lack of confidence in his entire organization. In all seriousness, that comment profoundly affected me. It would be tragic and irresponsible to tell the boss I had to catch a plane. This boss started a conversation at the end of the meeting. This conversation was important. This conversation was about the future. This conversation is a senior leader's ability to lead workers doing high-risk work successfully.

I knew at that moment we have not done enough thinking about why Fatalities and Serious Events happen within our organizations. The assumption is we had a lack of prevention. The reality is we had a lack of ability to manage the outcome of a systems collapse.

A fatality or catastrophic event is the worst possible thing that can happen in an organization. Nothing worse could happen. If you have had an experience with this type of event, you know exactly what I mean when I say, "the worst possible outcome imaginable became true." If you haven't had any experience with catastrophic events, it does not take much imagination to place yourself in this awful position.

Fatalities are the worst possible surprise.

There is nothing you can do about a fatality. You can't move back in time to correct the problem or change the negative outcome. There are no near misses to fatal events and serious incidents. You can't make this horrible story have a different ending. There is not enough money; no matter how much money you have, to buy a better ending to this event. Everything about your organization will be different, at least for a while. You will spend a great amount of money and time, money that cannot go to other priorities and problems. Customers will not feel the same way about your organization. Your organization's confidence will be rocked, seriously eroded, and that once strong confidence that allowed you to perform high-consequence will take a good while to recover.

One of your workers, one of your friends, a member of your work-family is violently removed from not just your organization, but also from life itself. A person was killed at your workplace for the simple act of doing his or her job. No warning. No time to prepare. No chance to stop the tragic event after it happened. No chance to prevent. Your worker is dead. Your organization is in collective-shock. You have no control or power; you have moved from magnanimous manager taking care of business to the kind of manager that permits circumstances in which people die.

Your organization failed to actively prevent a person from dying. Your organization did not have the power to stop a catastrophic event from becoming a catastrophic event. You are angry, sad, surprised, disappointed, and frightened. Mostly, you don't know what to do to make this type of event never happen again.

Fatalities and catastrophic events are terrible. We want to know what happened. We must know who or what failed. We have to know how this could possibly have happened without some type of warning or indicator that would have allowed us to detect and correct this terrible outcome.

I was interviewing a pilot for my podcast. The pilot was involved in an especially interesting commercial plane crash a couple of years ago. The crash was interesting, as was the interview, because of the complete lack of fatalities. This crash was the first complete "totaling" of a Boeing 777. The plane was completely destroyed, and yet not a soul was even seriously injured. During the interview, the pilot stated that at a certain point during the crash, he realized he had transitioned from a pilot of the airplane to a passenger of a plane that is crashing. Think of this? At one moment you are in charge of the direction and outcome of a flight and at the next moment there is nothing you can do but ride the crash out to whatever conclusion happens.

I can't think of a better, more effective, description of the fear and seriousness that must happen when your organization's systems have a sudden and complete loss of operational control and stability. It scares me. It also made me curious. What are we missing? How is it possible this outcome happens to an organization that has struggled every day to become even safer? What could we be doing wrong? What are we doing right?

What do we call these events?

SIF's, FSI's, Fatality Prevention Programs, High-Consequence Loss Reduction, Low-probability-High-consequence Events and even Highly Reliable Systems--Organizations are all part of the current nomenclature for these catastrophic events.

All of these titles refer to the type of event where people die and the facility suffers uncontrolled harm.

You may also notice that I am trying to use several titles for these types of events throughout our discussion. I know what we call them will one day matter, language matters greatly in our world. I am not convinced it matters what name we eventually will use for these events, specifically. I don't know that I care what we call these events. I care mostly how we see and think about these types of events. What we call these events is not nearly important as the need to see these events as special and different.

That said, I think the current multitude of different names for these serious-outcome events is a good sign. The many names that professional safety and reliability people are slowly circling around in order to settle on the perfect title means that we are struggling to understand these events in a different way, using different thinking. When we settle on a name for these events, the new name will have power all its own. That is what will matter most.

Perhaps over time, if we can make a case that is strong enough so these types of events will get a standardized and accepted name that will automatically trigger an entirely different risk assessment and control program. Then what we call these horrible-outcome events will only matter because the name will immediately create its own powerful response.

Chapter 3

Fatalities are not Normal

"I'll never understand how this happened."

I have heard these exact words spoken at so many fatality and serious event meetings with management teams. "I'll never understand – he should know better – he is one of my best workers." That mystery in how a good worker could be killed in the worst possible way is an important opportunity to learn. Not knowing how a good worker dies is a solemn and sacred opportunity to understand how to not fix the last event, but in fact fix the next event and the next and the next. We must be mindful that the seeds of tomorrow's accidents are sown today.

Perhaps one of the most difficult parts of my exposure to fatal and serious workplace accidents was not the profound sadness you find when an organization has a failure of this magnitude. It is terrible; the worst news an organization ever has to deliver. What are most difficult are the conversations the organization has with itself to desperately focus on "who did what wrong." After all, a person is dead or seriously injured; someone had to do something terribly wrong for this outcome to be possible. Really good people tell stories of how they missed something, or something they knew to be wrong had just gotten away from their control for a brief minute. They tell stories of how good workers became momentarily incompetent and could not manage the high-risk work. There are fears and sorrows so deep that the organization seems almost incapacitated in its ability to continue to do work.

There is an almost desperate need to explain how this worker failed to do the things that must be done in order to not get a tragic outcome.

This need is powerful and creates an environment where an organization's leaders cannot not think that if we could only roll the time back, we could change this outcome immediately. Our language and thinking begins to become focused on changing the past. We want to know why we failed to predict this event and therefore failed to prevent this event. In a way, we think a better organization would not have this outcome and this outcome must make us a bad organization filled with bad people who have bad outcomes. It is hard to not see this problem as endemic and predictive.

I want to emphasize three important learning outcomes from other organizations that have traveled this terrible path before to all who care to listen. In this case, learning from others is not only important, this learning is vital, life saving, experience:

1. Fatalities are not normal. Mostly people do not die while working at your facility. Most workers will spend their entire working career not getting seriously injured or killed. Don't think about a fatality as a failure of your organization's safety and reliability program. A fatality is an outlier event. Your organization does safe and stable work most of the time. This statement is just a fact.

2. Every minute you spend trying to establish who did something wrong, or poorly, or not hard or good enough is a minute of this almost sacred and serious response time you have wasted and will not get back. You are fogging your ability to learn by your emotional (and sometimes corporate) need to determine who is at fault. It may feel satisfying to determine who screwed up, but it is neither important nor informative.

3. It is not important to be right. Being right does not lead to new thinking. Being right only reinforces old thinking. Being right is not interesting. Being right does not create new learning. Being right feels important but could not be farther from helpful. Being right hurts your organization's ability to learn and improve. Being right is not effective as a retrospective tool. In fact being right does nothing to make your organization safer

and more stable. Being right tears your organization in two, which is not your goal and is certainly not an effective outcome. You will always be presented with multiple, over-lapping, conflicting stories from accident participants. Your job is to not assess who is right or wrong. Your job is to explain the event and tell the story of what happened in order to position your organization to learn and improve.

Your organization, and conversely the safety and reliability functions within your organization, is fixated on not having serious outcomes. You really do care and work your hardest to not have bad things happen. You don't have failure because you deserve to have failures. Something or things that were quite abnormal happened and those things led to a catastrophic outcome.

Fatalities and serious events are not normal. These types of events happen as outliers to the way normal work happens. These events are non-linear, non-predictive. These types of events are anything but normal accidents.

See fatalities and Serious Injuries as Unique Unwanted Outcomes, Outliers.

A fatality is not the logical ending of a bad safety program. A fatality is not the unwanted journey to the top of the Accident Pyramid. Fatalities are not normal. If we see the unwanted outcome as unique, our response...the same unwanted outcome...will be, by definition, tailored to that event – a specific and unique response. I know this sounds wrong...I know every bone in your body wants to ensure this event never happens again...and you're right. However, if you see this as a normal failure – the result of a chain of events or worse still the logical climbing of the accident pyramid – you will tend to work harder at the lower level events.

Working harder under the guise that the same motions that solve low-level events will also solve the more significant events is wasted motion.

Your organization will get better at the low-level events (you will have fewer ankle sprains) but you will not get more reliable around the high-consequence events, the fatalities and serious accidents.

In other words you want to shift your organization's thinking from solving the "failure to prevent the last event" to solving for a more effective matching of controls and safeguards for the next event discussion. Solving the event that just happened does not help understand or prevent another fatality event. I can't emphasize enough the need to build a body of knowledge around these catastrophic events that is purposefully different from classic industrial safety events. These types of events are not simple events that we failed to prevent. These types of events are complex failures that happen on many levels of the organization's operations and simply defy the ability to predict by the organization. Remember, when someone dies in an organization, not every system in the organization has failed, just enough of the organization's systems failed to allow the catastrophic outcome. It is valuable to know what systems failed and also know what systems worked in order to keep the bad outcome from becoming a much worse outcome.

As crazy as this may sound, this is a different approach to thinking about fatalities. In many ways the Sarah Jones case was possible because of the systems this crew used to do this type of work. They did fail to prevent this accident to be sure. They also were completely unprepared for this failure when it happened. As near as I can tell, those are two different statements about the same event. That is the discussion we have to have with ourselves first, then our organizations.

The case we are building is a fatality. It is not like other accidents. A fatality or serious event is much more like an outlier. Nassim Taleb calls them black swan events. Although in writing this, I know if you are in Western Australia a black swan is not rare; it is quite normal – let's put that little fact aside for the moment and get back to Taleb's central theme. Taleb's central idea states these types of events are rare and unpredictable events. When these rare events happen, we tend to quickly assume the event happened because we failed to prevent the event from happening, short and sweet, but this is the most wrong way to think about and respond to an organizational catastrophe.

The Black Swan: The Impact of the Highly Improbable is a book by the scholar and statistician Nassim Nicholas Taleb. Random House published his book *The Black Swan* in April 2007. *The Black Swan* focuses on the fatal or catastrophic impact of certain kinds of rare and unpredictable events (outliers or anomalies) and the normal and human tendency to find simplistic explanations or reasons for these events after the event has happened. This theory has since become known as the Black Swan Theory. Remember the power of retrospect in providing clarity and simplicity. Also remember that retrospect almost always simplifies the story and softens the complexity of an event.

Sound familiar? It should.

We have spent a lot of time presenting this one idea. Fatalities are not like normal events. Fatalities are not normal accidents. Fatalities are not normal and yet they are produced by the systems that don't usually make them. Odd in a way your organization's systems, which are normally stable and predictable, always have the power to produce an event that is not normal. Yet this event originates from the same stable systems and processes that normally do not produce bad outcomes.

Did I lose you in the last paragraph? To really discuss this phenomenon we must probably talk a little bit about statistics. Not a topic most safety people are raring to discuss, but I have complete confidence you are interested in this idea. Let's talk about anomalies. I mean really talk about how unpredicted outcomes are produced by stable systems.

Chapter 4

Fatalities and Serious Incidents are Anomalies

"Stop ignoring the outliers."

Let's define some terms in order to better discuss this next series of thoughts.

Don't get too worried. I will try to not bog this down in too many details, but the notion of an anomaly is important to understanding my basic premise: Fatalities are not failures of safety like years of traditional safety and reliability safety theory and training wants us to believe and rely upon. Fatalities are outliers; they are anomalies created by the very system that is trying to prevent fatalities from happening.

A statistical anomaly is the degree to which something (an event, a person, a data point) falls out of a normal range for one group (work process X) - but at the same time the anomaly is a result of being in that same group (work process X).

In other words an anomaly is when something weird happens in a task where normally nothing weird happens. This is important to understanding how a tree-trimmer can trim trees for 25 years and never get hurt, and then one Tuesday a tree falls and kills him. The tree trimmer is trimming trees the way he always trims trees and yet the outcome is completely different for this iteration of the tree-trimming task. The process that had the anomaly created the anomaly. Get comfortable with this definition; you will see it again in this chapter.

Fatalities and serious events almost never happen. Fatalities and serious events, even more accurately, don't happen until they happen. Fatalities and serious events are rare, Black Swan occurrences. This is not only good news, but also important to our discussion about prevention and response. Fatal events happen in systems that don't normally have fatal events. Fatal events happen in stable systems. (Fatal events probably also happen in unstable systems, but it strikes me if the system is unstable, we would work hard to bring the system back into statistical control.) To understand why this is important, we must first understand some mathematics and statistical theories. To understand the idea of rare events we must first understand the idea and theory of statistical anomalies in a stable data set, process, or system.

Lets reread the definition of anomaly given for this discussion: **A statistical anomaly** is the degree to which something (an event, a person, a data point) falls out of a normal range for one group - but at the same time the anomaly is a result of being in that same group.

Think of a statistical anomaly like this. You gather a bunch of information about some process or activity your organization does, whatever this activity might be; you use this information to understand and predict the outcomes of all the future iterations of that activity. After you have gathered this information for a while you learn that the outcomes, if you do the process the same way every time, are predictable. These outcomes are so predictable you can begin to categorize and expect the same outcomes every time the process happens. Amazingly this data prediction tool is almost always right on target. You can use this data to see into the future and good organizations are good at predicting the future.

You preform this now stable and measured process, this understandable and predictable process, and suddenly – out of nowhere - you get an outcome you do not expect. The outcome does not meet your normal expectation or fit in your normal categories. Yet the process was done the same way and most all of the other work outcomes are in fact the same category and same expected outcome. Nothing changed, nothing happened and yet you have a new piece of information you did not expect.

This new piece of information does not fit into your expected result or your normal category. You got a rare and unexpected result. This new outcome is not normal, yet you did the work in a normal way, using the normal processes. This new outcome is an **anomaly**. The outcome is real, and the organization used the exact same process; yet, the outcome produced by the process is not the expected outcome. This outlier, this odd result, is an anomaly.

Remember our definition: **A statistical anomaly** is the degree to which something (an event, a person, a data point) falls out of a normal range for one group - but at the same time the anomaly is a result of being in that same group.

If you flip a coin 100 times and you learn by carefully recording the flipping outcome of each flip (pun intended) that the coin lands at a rate of about 50 percent. 50 percent heads and 50 percent tails. This result is stable and predictable. Every time you flip a coin there is a 50 percent chance it will land either heads or tails. Your process is mature because your results are predictable. Every time you flip your coin the outcome is as expected.

The following week you flip a coin 100 times again using the same process you always use to flip your coins, only this time the coin lands on its edge. Holy Cow! The coin landed on its edge. That is neither a heads landing nor a tails landing. The coin landed to be certain; the coin just did not land in a way you have become accustomed for the coin to land. The edge landing of the coin is really unusual and rare. The edge landing is not impossible because it just happened. The edge landing is, however, highly improbable. In fact the edge landing will be almost unbelievable. My guess is you would take a picture and have a meeting to talk about the day the coin landed on its edge.

That edge landing is an anomaly. **A statistical anomaly** is the degree to which something (an event, a person, a data point) falls out of a normal range for one group - but at the same time the anomaly is a result of being in that same group.

The alternative would be if you ran the same coin flipping process and your results were an 80 percent heads and 20 percent tails. Again that is a different process outcome of the expected 50/50 percent. What you expected to happen did not happen. That also would be some type of anomaly, although not as severe or as interesting as the edge landing. In fact, your organization would probably study in great detail why the percentage of landings had changed. Perhaps you would be able to trend this new data (wind speed or table surface) in order to manage the unexpected outcome. This data would be valuable to your organization's future – if your future depended on a 50/50 flipping average.

The crazy thing is the system that "caused" the anomaly is also the system that creates the stability. I can't seemingly say this enough. I have defined this term four times in this chapter. This attribute, stable to outlier, is so vital in understanding fatalities and serious events. The fact that the same system that creates success has the potential to create fatalities and catastrophic outcomes. Yikes! By all rights that should freak us out – however mathematicians have known this for a long, long time. Anomalies are neither normal nor abnormal – they are rare events. I guess everything and nothing is possible in math. Listen to the sound of one hand multiplying.

We can't manage anomalies by doing the processes and work better. We can't manage anomalies by using previous data to predict the outlier-filled future of our organization. We can do the processes; follow the rules and procedures, as carefully as is humanly possible and we cannot prevent anomalies from happening – at least not with any guaranteed result. We simply know an anomaly is not predictable, but probable. You can't really say the new and odd outcome never happens – because it happened. You can say when you flip a coin the chance of it landing on its side is extremely rare, almost impossible to imagine happening, but it happened.

Fatalities are not normal. Fatalities don't happen all the time. Fatalities are incredibly rare in organizations (thank God), and yet fatalities happen. It is easy to see a fatality or serious incident as a complete failure of an organization's safety management system; however, I am not certain placing blame on the safety function of an organization makes much sense. I could make a convincing argument the safety system did not fail the operations of an organization. I could make a case that says the operations of an organization failed to operate safely. Placing blame on operations is not much better.

It is probably better to think of fatalities and serious events as an outlier event – an event that simply is beyond the ability for a safety program to be successful in managing the outcome of a major event. This is fairly important, because any organization must get around the need to find a reason, a bad apple, for how such a terrible event could happen in our organization. The power of retrospect to simplify the event is strong and rational. When you figure out as an organization this bad outcome is rare and falls beyond our ability to manage with our existing system, an event beyond the safety basis, the opportunity to learn and improve because of the event moves to the forefront. You have also moved beyond the simple answer to a more complete and context-rich beginning of the solution.

High-Consequence, Low-Probability events are not the enemy to be battled, but instead a factor to be controlled. We can't stop anomalies, but we can control an anomaly's outcome. Think about that for a while. You will never have an organization with zero anomalies. You would never even try to make such a claim. A claim of zero outliers is going against the fundamental mathematics of any process.

I recently was a part of a meeting where the main premise around the prevention of fatalities and serious events was centered on critical decision-making. It sounded good, but what was really happening was a "blame shift" from worker to leader. I am completely convinced this tactic is not better, and in fact much worse. Our need to establish a worker (or leader) failed to make a critical decision that would have led to a better outcome is strong, but inaccurate. Fatalities and serous events are not about confirming how the old prevention methods failed; they are about understanding how the new system was pushed beyond its capacity to recover. Be cautious about blame; it is an easy way out intellectually, but not sustainable for your organization's future.

We are arguing with a powerful, cultural myth, the idea that somehow these events are the ultimate failure of our safety systems. This simply does not seem to be true based upon the evidence. We somehow desperately want there to be a connection between the frequency of events and the seriousness of events. All our observed data does not hold this connection to be based in real operational experience. You do not need to have a whole number of smaller events to have a catastrophic failure. The other side of that argument is that if you have a lot of smaller events, this doesn't necessarily point to an impending fatality. Outlier events do not normally have low-level indicators. In this one case, and I say this in this one case, the small signals are not predictive.

You can see the years of the Safety Pyramid have left a deep scar in our thinking. People often refuse to believe the Pyramid is not correct. There is a desperate need to hold on to the elegance and predictability of the Pyramid. I get it. I, too, want the world to be predictable and linear. I want everybody to happy and rich. The strong desire for predictability does not create predictability. Believing in the Pyramid would make it possible to have zero anomalies. Sadly, this fact is not the case.

Chapter 5

How Fatalities and Serious Events Differ from Other Events

It probably should go without saying, but Fatalities and serious events are not like other accidents. I realize this is painfully obvious to any organization that has had to endure the pain and recovery of an event. The difference between fatalities and serious events is about so much more then simple, yet horrible consequence. We must understand a bad event is not a bad event because the outcome is especially bad. Fatalities and serious events are different because the tools we use to manage and mitigate the classic industrial safety events in our organizations did not work for these events. This is a not a super accident, like a giant ankle sprain; this event is catastrophic. These events are giant consequence events that do not rely on other smaller events to happen. You will not predict a fatality or serious event, and yet you predict them all the time. This is truly operating in the middle of a paradox. Get used to this fact. It is a big part of moving forward in this thinking.

However obvious, the difference between the two events must be explored. Our traditional view of events has not allowed us to think of these events as different. In reality, we take a graded approach to accidents. We believe big accidents will have bigger causes and small accidents will have smaller causes. We know big and little cause is not an intelligent way to view events, and we know fatalities are not total failures of our systems – we simply have been led to believe these types of events are the upper-limit of the graded approach model. That's not right, so let's think about why.

When something unwanted happens in our organization, we immediately assume that something important must have failed in order to cause this bad event to happen. We then tend to investigate the event and look deeply for some weak or non-present "thing" that should have either happened (follow procedures of stop work) or should not have happened (short cut or unapproved activity or tool). Most every event I review generally follows this above-mentioned model. When we find the "thing" that either happened or did not happen, we then begin to do our analysis.

Traditionally, our analysis activities have spent much time looking at the event that happened in order to identify the thing or things that failed — and then we go out and fix those things. We spend a lot of time, energy, and effort looking deeply for a "root cause" for an event that happens in our organization. We have been trained to believe that an accident happens because things failed and caused a bad outcome to a person, product, or process. Failure is the logical result of a series of bad choices and/or bad systems aligning in such a fashion that the bad thing that happened-became possible.

Somehow our organization allowed the outcome to happen. We became weak, complacent, or inattentive. The problem now, knowing we are bad, becomes creating a future where this outcome that just happened is prevented from ever happening again. We are destined to believe if we had been better at prevention we would not have suffered this terrible outcome.

Fatalities and catastrophic events are not like other events. Other, less serious events seem to happen because we missed something that could possibly have prevented the event from being successful. We missed a signal in our system that would have indicated something of consequence was about to happen. We could have stopped this event from happening by simply being better or more aware of our operations happening inside our organization.

Fatalities and serious events seem to "shy away" from smaller signals. There probably are not very many near misses that would indicate a fatality is about to happen if a fatality is an outlier to a stable system. All of the efforts that you and your organization have undertaken to not kill people – all that prevention thinking you have done – is probably never going to be sufficient to stop *all* catastrophic failure.

You must prevent events from happening. You will never be at a place where you would stop doing hazard identification and condition assessments. These actions are vital to the work you do in keeping your systems safe and reliable. Because fatalities and serious events are not predictive, you must learn to expand the way you do these assessment to both identify prevention issues and the controls and safeguards that must be present to do this work safely.

You would not want to drive a car without seatbelts. Although the chances you will wreck are slim, the recovery that the seatbelts provide seems fundamentally important to your safety. This is even stronger when you put your family in the car with you. You have already learned you have to manage event probability and event recovery in parallel. This same thinking applies to fatalities and serious events. You manage both the probability and the consequence for this event at the same time.

Prevention works and works well. Your prevention activities should not stop. Your prevention activities are keeping people from having accidents. Prevention is key, foundational, to a strong safety and reliability program. Prevention is not enough. Prevention can't handle consequence. Prevention tends to be a seductive way to imagine the organization's ability to predict the future. Prevention can do a lot – does a lot; prevention just cannot do it all. The problem isn't really prevention; the problem is that prevention depends on prediction and accurate and effective prediction is difficult.

Effective prevention is an outcome of effective prediction, and prediction is difficult

Before we get too far into this conversation about serious events, we have to discuss the powerful meaning the word "prevention" carries among organizational leaders. Saying serious events are not preventable brings an emotional response to some leaders. In a way, talking badly about prevention is taking one step too close to the most sacred of all sacred safety cows, the belief that all accidents are preventable.

I don't mean to downplay the importance of prevention. I do mean to challenge the way we think about prevention as the most important, most effective, and most powerful reliability tool an organization has. Prevention is not completely effective or completely powerful and sadly not the most important tactic an organization has to build safe and reliable operations. Prevention seems like a great idea, prevention does not seem to deliver on the great idea that it promises. Most importantly, prevention is not prediction; these two concepts are vastly different.

Predictable events are easy to prevent. Every organization will do everything in its power to stop a serious or fatal event if the organization knows that the event is going to happen. The challenge is not in the word prevention, the challenge is in the ability for an organization to predict a serious or catastrophic event in order to somehow change operations in order to prevent a serious outcome.

Buckle up for the rest of this conversation, but don't get overly concerned that the discussion takes aim at prevention. Prevention is important, but not sufficient in and of itself to manage catastrophic events. This discussion is what follows.

Chapter 6

There is a Bias to Prevention

Let's state the obvious:

> *Every bad thing that happens – happens because we failed to prevent it from happening. This statement is always true...but is this statement ever relevant?*

Prevention is attractive and important. Prevention keeps a messy world from becoming messy. Prevention is so vital to our overall organization's success. Prevention has, over time, grown to be a multi-billion dollar industry. We really want to prevent everything bad from happening. So much so that we have, over time, become overly reliant on our prevention and planning efforts. This overwhelming belief that prevention is the key has created a gap in our readiness for failure.

Remember this: Prevention efforts can't prevent causes that were not expected. Just as planning can't plan for unexpected events. Prevention can't prevent anomalies in our normally stable systems. You, by this time I am sure, are probably starting to realize prevention is not a control; prevention is not a safeguard against the consequence of a significant event. There is a real bias towards the belief when something bad happens, it happens because prevention failed. This bias, I am afraid, has caused much harm. What scares me is the harm this bias may continue to cause in the future.

We are so invested in preventing all accidents we don't build the ability for our systems to recover if the accident happened. In reality, this bias is even worse then we all may fear in that most event learning is actually done in order to reinforce this bias. We actually investigate to determine how we failed to prevent the bad thing from happening.

We have been taught our whole working careers all events are preventable. I would go as far as saying many of us (and many of our workers) truly believe every accident could have been, and should have been prevented. I think this idea is wrong. I don't believe all accidents are preventable.

I don't think the people who believe every accident is preventable are bad people; clearly we have trained and over-trained this idea to entire rooms filled with students. Where we have made the most impact with this message, and therefore will have the most difficult time reigning this idea back in to our control is with our middle managers. These folks and others really do believe every accident is preventable. *You* may believe all accidents are preventable. With as much compassion as I can muster, I boldly ask you to let go of this non-empirical safety myth. The assertion that all accidents are preventable is simply not true and not possible. Accidents by definition are accidents. Holding dearly to false accident prevention beliefs are not value metrics for people...accident prevention beliefs are a safety-marketing tool that has lasted beyond its purpose.

These bad events that happen are accidents. These events are unintentional deviations from an expected outcome (that is a moderately fair definition of an accident). We did not know, nor could we have known while the accident was happening, that the accident would happen. The likelihood of a bad outcome was the furthest thing from our minds. This event was an accident. By definition an accident is not preventable. Sorry guys, I know this is a difficult thing to read.

This is a sacred cow in safety--this belief we must prevent all events from happening before these events happen. The belief is not wrong thinking; in fact, this thinking is probably the morally right way to think about managing high-risk work. It just feels bad to say our systems and workers will fail. This belief is simply not a control or barrier to *actual* prevention. So when prevention doesn't work it doesn't' work – prevention is not enough. I doubt you would fly in a commercial plane that only uses prevention as a strategy for reliable operations. I know I would not. I want a plane with a secondary hydraulic system.

The real problem is the sacred cow belief "all accidents are preventable." It colors the way we think about accidents. We see bad outcomes as the product of failure to prevent the bad outcome.

When something bad happens, it happens because we failed to prevent the accident. When something bad happens, it happens because our bias is based on why we did not prevent the bad outcome, not how we responded when the bad thing happened in order to reduce the negative outcome.

This is sometimes controversial. I don't want to be so naïve as to not realize the idea that prevention programs probably are a large part the problem and are not the solution to stopping or removing catastrophic failures. I know people may think this idea is wrong, even immoral. It is not. In fact, in my experience, organizations that experience the death of their workers often have among the best prevention and work planning programs in the world.

We have to get over the "failed to prevent" mindset

We can't prevent fatalities...or worse still...we don't prevent fatalities. Yet, we prevent fatalities every day in our organizations.

Boy, that sentence really makes people angry and defensive. I understand why they feel so harmed by the statement fatalities are beyond prevention – but I don't understand why professional, thoughtful people are so ill prepared to think differently about this type of event. I really mean this statement. I have had some smart and important people just shut down when confronted with this idea.

This is not an argument about moral failing. I am not making the case that we should not prevent serious and fatal events. I am saying that if we fail to predict we will fail to prevent a catastrophic loss of any type. Deep in this discussion, lies hidden some type of idea that we must prevent all catastrophic loss – an idea that is simply made impossible by our failure to predict or imagine all catastrophic loss.

In a way the failure to prevent argument is a bit easier when you are working with a facility that has experienced the death of a worker. That facility really is able to access some urgency and importance in changing the way they think about doing safe work. It is easy (and a bit cheap) to simply say we should have prevented this accident from happening. Of course that is right; you should have prevented it from happening. The problem becomes the almost myopic need to then investigate why you did not prevent the accident. That thinking does not help you get better.

One of the first challenges is going to be preparing the soil to plant and grow this different way of thinking. We cannot expect the same way of thinking to get us to some type of better outcome then we currently have. In many ways, the way we define fatalities as a problem will absolutely color how we create new and useful solutions to this problem

Let's talk about how we have traditionally seen these types of events:

Where once we saw a fatality as the logical outcome of a bad safety program, – the safety program did not fail the worker – somehow the worker failed the safety program. We are now starting to see quite the opposite in our understanding of a catastrophic event. The safety program, most likely did not fail operations – operations, most likely fail the safety program.

This shift in thinking carries across all the "closely held beliefs" we have for the safety and reliability profession. Where once we saw workers as the problem, we are now starting to understand workers provide a constant detection and correction function to our organizations systems. Where once we saw a fatality as the ultimate accident, we are now starting to see a fatality as an operational outlier. Perhaps the biggest shift in thinking is the way we think about cause. Where we once thought the worker caused the fatality, we now are starting to think the worker was not the cause, but in fact the "trigger" for the events to happen the way the events happen. These shifts in thinking are happening. You probably can't stop this action, and they are important and valuable in shifting thinking about our future programs and success.

A Shift in Thinking

Industrial Safety	FSI
• Classic Prevention Model	• Outlier Events
• Seen as predictable	• Surprise Outcome
• Seen as preventable	• Not classically preventable
• Manage low-level indicators	• No/Few low-level indicators
• Near-miss data	• No near-misses
• Individual failure	• Operational failure
• Individual Intervention	• System Intervention
• Worker seen as the problem	• Control problems
• Owned by Safety Effort	• Owned by Operations
• Fix Worker Judgment	• Manage Consequence
• Faith in system	• System not understood

Fatalities are not a failure of prevention. Fatalities are a failure to control the serious and unwanted outcome. That difference is important – but I fear may be seen as a difference in language and not a difference in thinking.

Believing All Accidents Are Preventable is Ridiculous – and Dangerous

Let me set the stage for little conversation I had several years ago. I am in a fancy-pants meeting room filled with important people. I am talking about the "New View" of safety and these folks are just not buying what I am selling. I whip the crowd to some type of crescendo and try to really drill the idea that all you can manage are your safeguards and this conversation takes an odd turn...

> *"This is all fine and good, but lets agree on one important fact...All accidents are preventable,"* says the man in the nice shoes sitting at the head of the table.
>
> *I respond, like I always respond by saying, "That is only true in retrospect...every accident is preventable after it happens...the problem is that accidents are not predictable or imaginable before they happen. That is why they are called accidents."*
>
> *"I just think we will have to agree to disagree,"* the nice man says, completely discounting the last three hours of discussion in which this group had participated.

Let me say this about that. Saying we "agree to disagree" is intellectually lazy, goofy, and really scary. We don't talk about safety in order to agree to disagree. We talk about safety to understand that reliability of a system is a capacity that is designed and maintained within the system. We talk about safety in order to build a case for safeguards and recoverability. We talk about safety in order to help organizations understand you don't drive human error out of a system – you build human error in to a system.

The idea events are not predictive based upon severity is why we don't "agree to disagree." This conversation is about human lives, and how we help people fail safely. The conversation is about doing work in a constantly change work environment while maintaining the margins to fail. That is what this discussion is all about.

But what can we do with ideas like Heinrich's Pyramid, the notion every accident is preventable. We have made "every accident is preventable" almost like an oath and we have talked our people into believing and repeating this statement over and over as if calling up the power of the statement. I have news for the world. The statement "every accident is preventable" is not based in reality. It is a moral statement, usually made after the fact, and is really quite an elaborate way to blame a worker for failing to prevent the event.

So let's deconstruct this idea a bit. My guess is this statement was originally made in order to empower workers; the folks at the sharp end of the work stick, to believe they can prevent bad outcomes from happening. I will give the idea the benefit of the doubt. I am not arguing the moral notion that accidents can be prevented. I am quite convinced we prevent many, many accidents. I am also incredibly convinced prevention is not solely sufficient. We are doing the wrong thing and we are not doing enough of it.

Safe organizations kill workers...

It is all about the bias. We tell ourselves we must prevent every accident, every time. In fact, we believe we live in a world where not preventing an accident is criminal.

If we believe we can prevent every event, we don't really need to have the ability to use controls and barriers to manage event outcomes. If you don't believe you have a potential weakness, you won't believe you need to prepare the organization to fail safely.

The preventability bias gave a lot of fuel to the zero goal issue in safety management. If you truly believe all accidents are preventable, then it is possible to believe our operational target could be zero accidents. This crazy zero accident argument has really perplexed workers and leaders since the zero outcome idea became socialized in our organizations.

Zero is Killing People

This is the most interesting artifact of the old safety thinking. Zero seems like such a perfect goal for safety. Zero accidents, Zero harm, Zero injuries, Zero events, all of these ideas sound so good and so attractive. All of these zero's are simply not possible and will never be possible. Your organization places its entire future in the hands of smart, yet unreliable human operators. You will not get to zero. You will never sustain zero. Asking for zero actually reduces your operational knowledge. If you want zero events – I can almost guarantee greatly reduced reporting of events of all sizes and shapes.

You have heard it. You have said it. Zero accidents are attainable. This idea is just so wrong and yet feels so right. The problem with this zero idea is that zero reports feeds the myth that accidents can be prevented if we were simply better at prevention of accidents. This is a really dangerous logic loop into which we can fall, deeply. If something like an event happens then we failed to prevent the event from happing and will not get to our goal of zero. Therefore we must be better at prevention of accidents in order to get to the goal of zero.

Zero makes the mathematics hard. Any small event that could potentially become recordable will take you away from your goal. You are asking people to hide data that could potentially be incredibly valuable to your organization. If I fall down in your parking lot and no one sees me fall and I don't get injured, did I fall? Would a zero culture encourage the reporting of that event? No, the answer is that a zero culture would rather not talk about events that are quasi-reportable, I fell in your parking lot, the event happened, but because I was not injured and nobody saw me do it, it is really not an event. This gets even more complicated when you couple zero safety reporting with "cardinal rules." If you are going to fire me, I will make sure your number is zero, even if I bleed in my car all the way home.

See the problem? It is as if the failure to prevent the accident was some type of conscious and deliberate choice. Accidents happen because we purposely choose to do something other than prevent the accident we had from happening. Really? Is that what your organization really thinks? The answer is "no," or at least I hope the answer is "no."

It will be hard to "put this toothpaste back in the tube," but I would strongly advise you rethink your zero programs. You don't have to admit that the thinking for the zero program concept was well meaning but was wrong--that would be really painful – but your workers would probably agree, you can say that your definition for safety has matured and grown with your work complexity and technological advancements. We really see safety now as the presence of capacity, not the absence of events. Lots of books, include some that I have written, discuss this in deep detail. If you have not exposed yourself to these ideas, it might be a good idea to stop now and grab a "new view" safety book and give it a read.

For all the reasons discussed you should really rethink your zero program, but all of these above reason are not what this discussion is about. Let's talk about the operational bias that zero creates in the way your organization, every level of your organization from the top brass to the lowly temporary laborer, thinks about events like catastrophic failures and fatalities.

Zero reinforces the belief that we failed to be a zero organization. We failed to prevent the accidents and all accidents are preventable. That bias towards prevention is reinforced in your organization's language and systems. In many ways your organization is getting the exact outcomes it does not desire. We have to think about how zero influences our operational thinking, planning, and understanding. We are destined to fail if we start with an impossible target. We are asking for something we don't want – reduced operational information. We don't know what we don't know because we don't want to know anything that isn't a zero.

Let's leave safety and reliability and think about this idea that a system must support itself in order to be effective. It seems like a ridiculous thing to have to say, but many of our organization's systems tend to create more goal conflict then these systems create operational data. Here is one example of a system that constantly has the ability to feed back in to its own operations to provide real-time data to keep the system effective, stable, and safe.

Case Study

Taxi Differently

I think I really like these new web-based ride share apps like UBER and Lyft. I can tell you that these new "shared ride" services have changed the way I travel more then most any other recent travel improvements.

That sounds like a rather weak endorsement, but it was such a different idea to me at first. My sincerest apologies to the taxi differently people. I know little about what it took to start this paradigm shift in public transit, but I stand in awe of how you made a program that is basically "taxi-differently." It took me awhile to get the guts to use it, but after the first ride I was hooked. It is easy, much more affordable, gives me a great receipt, and I never have to even touch my billfold. I wondered how UBER and Lyft have done what it has done so quickly and so effectively? The answer is in knowing what is happening, actually happening, all the time. The answer is in the real access to the operational context all the time.

In order to have real access to the actual operational context of your organization, you must know two things:

1. How work is actually done, in real time with observable data.
2. What your systems truly reinforce. In this case your system to reinforce the system outcome you want to have.

The burden to know is on the organization, not on the workers. In order to absolutely grasp the effective processes and the real conflicts that exist in your operations and systems, you must access the workers who do the work. Workers know which part of your process works and which parts of your processes don't function the way you believe (or was intended when designed) they function.

The system's reinforcement question is even better. To cover this idea let me give you an example I see all of time. I check in to lots of hotels. I am constantly amazed how much keyboarding the desk personnel have to do when I check in to the hotel. I know my data has been entered before I get to the hotel. I know this because I put most of this needed information into their system when I made the reservation.

I am constantly stunned by the amount of times the workers have to re-key this data in to their systems. A perfect hotel management system would only take two or three keystrokes to complete this transaction. This is an example of a system that does not support itself. The computer should reduce the worker interface, which has two benefits: less work for the worker and more accurate data handling for the organization.

This system idea is what makes me more and more attracted to these new types of taxis. I am really amazed at how these new companies create consistent product delivery and customer happiness, not by managing driver behavior, but in fact by creating a system that actually reinforces the operational outcomes that these companies needs in order to be sustainable and successful.

Every taxi differently ride a rider arranges is measured. UBER and Lyft know who opened their app and chose to take a ride, where that rider started and stopped the journey, pricing for that journey is based upon immediate demand, and the driver is paid based upon the successful completion of this journey. These new view service providers then allow the rider to rate the driver, which dramatically affects the driver's ability to get the next ride. UBER and Lyft also allow the driver to rate the rider, which creates a sense of social fairness for the driver. These companies measure everything and are constantly learning and giving feedback information in to their systems and operations. Don't believe me? It is worth a look to understand and be a bit jealous of how they monitor their performance reliability. UBER and Lyft use real-time, actual data to constantly attempt to achieve operational excellence. They don't assume operations are effective; they are constantly pulsing operations to adjust service, and to adjust availability, and most importantly adjust pricing based upon demand. Every part of their operations is assessed based upon real action – not on perceived (or ideal-state) action. These taxi differently companies know how work is actually done.

UBER and Lyft allow you as the rider to rate the driver. There is an incentive to be the best driver possible in order to keep your rating as high as possible. Higher ratings mean more riders and more riders' means more money. This incentivizes safety, efficiency, and overall service to the customer while allowing operations to be as efficient and effective as possible. This work control system is a brilliant use of peer-assessment and an excellent example of a system that has been designed to produce and support itself. These companies also allow the driver to rate you, the rider. That incentivizes the rider to be polite, clear, and well behaved as a customer. Again, the system supports a successful operational outcome.

All of this is based upon a business model that is constantly pulsing real operations. Accurate and real-time data is used to address demand and supply in real time. It is elegant, effective, and produces consistent outcomes in a complex operational environment, quite remarkable and worth a few moments of your thinking. Could your systems actually improve operations? Seems odd to have to ask these questions, but in my experience most systems are built to support the systems survival, not the operational outcome.

What would you say if I told you most companies that manage safety and reliability do this work with very limited, if any, data about the actual performance of the work? Would you think that seems ridiculous and foolish? Knowing less about operations in real-time does not allow for you to constantly adjust and align operations around need and risk. Our present structure actually incentivizes NOT knowing real-time information. Our present structure wants to manage the work in the ideal state – not in the actual reality, the current context, of what is happening on the shop floor.

The desperate need for predictability – The Never-Ending Hunt for Leading Data

So, I have an interesting question for you, "Why do we want work to happen as it was designed to happen and not as it is actually happening?" What is it that makes us think the procedure is the only way, the one right way, to do a task in a constantly changing and adapting work environment? That is an interesting series of questions.

My guess is it all has to do with a goofy pyramid, and dire need to control the future, and a quick and simple answer that seems to make everything better. What a long and powerful history this goofy pyramid has had.

That idea, a desperate idea, that somehow everything that happens in our facilities is the logical escalation up the safety pyramid is so harmful to our thinking and really harmful to our ability to learn and understand how work happens in our organization's operations. The goofy pyramid is reducing your operational data not increasing your operational knowledge. This fact alone is reason to move forever away from the goofy pyramid. That will be the last time I refer to the pyramid as goofy, but it was certainly fun while it lasted. Thanks for your indulgence.

You see we have been taught there is a hierarchy of harm in the safety industry. Heinrich told us safety happened in a pyramid and if we prevent little events the same actions we used to prevent the little events will prevent the big events. His assumption that frequency and severity were linked was elegant, seductive, and super predictable. The problem is this notion was wrong, really wrong and not supported by empirical evidence. In some high-risk, regulated industries like nuclear power our offshore production, risk is defined and calculated as probability *multiplied by* severity. Safety analysis engineers struggle to predict every possible combination of failure scenario in order to develop systems and strategies to prevent the failures imagined. It is a pretty good assumption this discussion will not make the traditional view of risk calculation people happy.

Herbert William Heinrich (1886 – 1962) was the Assistant Superintendent of the Engineering and Inspection Division of Travelers Insurance Company. This part of Heinrich's biography should give us a brief bit of warning that there will be a bias around the economics of accidents in Heinrich's thinking. Not that that is a bad approach-- accidents cost organizations much money, but we are starting to get a glimpse that the model will be economic in nature.

Heinrich published a book in the 1931; this was really the first academic reference to the Safety Pyramid. This book was called, **Industrial Accident Prevention: A Scientific Approach.** The book clearly did very well and must have sold a bunch of copies over the years, although to be honest I had to hunt long and hard to find a copy of the book to even try to get to a primary citation. One of the more attractive findings of Heinrich's book was the empirical (although now this is quite in question) 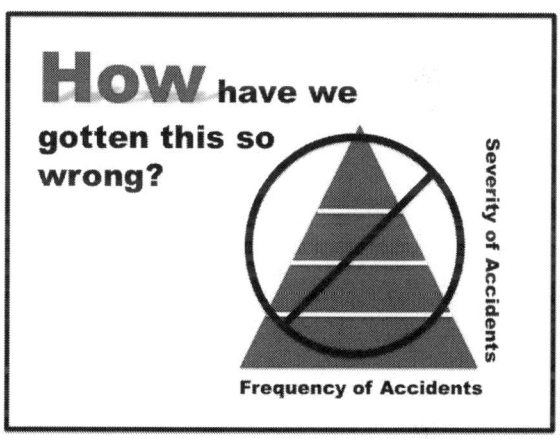 finding that in a workplace, for every accident that causes a major injury, there are 29 accidents that cause minor injuries and some 300 accidents that cause no injuries.

This was astounding to the safety and reliability community. The idea that accidents are that predictable was really earthshattering. Heinrich posits on this idea because many accidents share common root causes, addressing the more commonplace accidents that cause no injuries can and will prevent accidents that cause injuries up to and including fatalities. The same motion, actions, programs, tools, and behavior control mechanisms that stop the small events will stop the larger events.

And there it is, in all its glory and history, the origin of the "magical connection" between frequency of events and consequence of events. Because these concepts came from the insurance industry, who really were the early leaders in industrial safety, and because these ideas seemed to have an empirical basis in real observations of industrial work, this idea of solving small problems in order to avoid large problems got a lot of traction early.

Everett Rogers, the father of the theory of the Diffusion of Innovations, would call this idea a diffused concept that moved quickly along the diffusion curve. Rogers would say that this idea diffused quickly because the population wanted this information to be true in spite of the truth that was not present in the information. The desire to simplify and predict the future has always been compelling to human beings. Heinrich was simply the man with the solution, at the right place and the right time.

Heinrich came to his conclusions after reviewing thousands of accident reports that were completed by field personnel. Heinrich's data set would not be different from behavior observation cards that are completed in the field by field workers. Because of this method of collection for accident data, contemporary scholars believe there was a historic tendency to generally blame the workers for causing the accidents. Because these were historic field-based data collection tools there was little, if any, actual investigations done on the events.

In my opinion Heinrich's heart was in the right place. Sadly, Heinrich was seduced by the hopeful belief in the predictability of the outcome. Just like we, even today, are seduced by the idea there is some "leading data set" that will make all the difference in our safety and reliability programs. After all, if this were true, this would be an amazingly helpful thing for the industrial world to know. This safety "magical thinking" invades a lot of our safety programs even today, or should I say especially today.

Magical thinking is always a problem.

Believing in something that is not true may feel like the right thing to do, but that belief carries with it consequences. It feels good to ask workers to care more and try harder; however caring more and trying harder does not make your organization safer. Caring more and trying harder aren't controls that help manage hazards – they are just "magical words" that make the world a simpler place.

Probably Heinrich's most cited data would be his statement, "88 percent of all workplace accidents are caused by 'man-failure' " (his words not mine), is perhaps his most impact-fully wrong idea to industrial safety and reliability. This idea causes much harm to understanding actual catastrophic failure. To be fair, Heinrich did go on to tell employers that the managing of the workplace hazards was more effective then managing worker decisions and behaviors, but after the grand statement that almost all of the problem (88%) was of the workers doing, you can't really blame management for going after worker behavior. I am just so surprised Heinrich's ideas had the stickiness to last for almost 100 years in the mainstream lexicon of safety. I feel lucky if I can get a management team to remember something I said for two weeks.

Ultimately it is probably not fair to fault William Heinrich for all of our organizational and industrial bias's towards prevention – but he is not without sin in this story. Heinrich was clearly no innocent in his research methods and practices. His data set was never found, and therefore all the scholars that tried desperately to revisit his study could never really build a strong empirical case to prove his theories. My take on it is that we wanted this Safety Pyramid to be true so much that we simply never questioned the thinking around this idea. It probably helps to come from one of the largest insurance companies in the world at the time as well.

Our more contemporary problem now, almost 100 years later, is "wanting" an idea to be true is not science and does not make the idea true. The larger problem is this old and inaccurate thinking has influenced the way we think now, and have thought historically about fatalities. We really believe that if we were better at prevention we would never have serious events. If we were simply better at preventing fatalities we would not have any more fatalities. Remember earlier we discussed the artificial supply that is always created for real demands that are not being fulfilled? Artificial supply became what we now know as the safety pyramid. These facts that the safety pyramid claims to present and supported as predictive safety data are not based in actual fact.

The times, they are a changing

We have battled a couple of long-standing and significant safety myths in this little book. We took on the idea all accidents are preventable. We blasted the unrealistic and operational silencing idea zero is attainable. Finally, we have built a case that Heinrich's safety pyramid is based upon some data sets that simply are not capable of being repeated.

> **We have built our entire profession around the idea that we can prevent bad outcomes from happening...**

All three of these ideas are strong and powerful ideas in the traditional belief sets of safety and reliability professionals and leaders all over the world. All three of these ideas are also based upon a fundamental belief that failures in operations are actually a direct result of failing to prevent the accident from happening. We have built our entire profession around the idea we can prevent bad things from happening.

Pay attention to the power we have given our safety programs based upon the prevention bias we carry in our thinking and practice. In some ways, we have thought ourselves into a corner we must get out of before we have more serious events and fatalities. Not a simple endeavor, changing the way industry thinks about operational reliability, but an important task for us all.

The bias that all events are connected by frequency and severity is leading us towards fixing the wrong parts of our organization and leading us to the belief if we were better at prevention there would never be any hazards to control. This is the battle we must fight. Stop using pyramids to talk about safety! Stop it! They influence the way we think about the problem, which in turn, influences what we try to control and fix. Sadly, those are not the same things that provide the foundation of fatalities and serious events.

One final note, I know this is really an ingrained idea in our workplace safety programs. I also know it is almost easier to simply let this folklore continue on, unabated, because "what can it hurt?" It really is a dangerous way to think about managing high-risk work. It is dangerous because organizations don't pay attention to the more risky parts of their work because they believe their attention is best served, most effective, on the small-scale events. We really do work in places that spend more time on sprained ankles then they spend on confined spaces. That is just wrong on so many levels. Thousands of hours are spent investigating low-level events in order to stop fatalities. Those hours spent learning how ankles are sprained are really quiet wasted and do not lead to new thinking or better controls in our high-risk work.

The Prevention Paradox

I think every time you tell someone their shoe is untied you may have prevented them from falling down and breaking their arm. The problem is you don't know if the untied shoestring would have caused them to fall down. The untied lace certainly had the potential to cause a fall. However, the untied lace would not necessarily cause a person to fall down. It is hard to know if your warning to tie an untied lace is actually stopping a bad outcome. This argument is a bit "paunchy" in illustrating that the statement "all accidents are preventable" is a paradox. The statement is both true and false at the same time and depending on context (when the statement is used) could be either true or false.

In case it has been a while, my favorite definition of a paradox is the old standard that reads something like this: A thing that is made up of two opposite things and that seems impossible but is... Here is an old tried and true example of a paradoxical statement: "This statement is false." If the statement is true then the statement is false. If the statement is false then the statement is true. In a paradox context matters in order to understand the thinking of the statement. Digging too deeply in to a paradox will probably make you crazy, so don't do it too much. But think about our original argument, the one where the boss wants to "agree to disagree." That is a function of the paradox present in the statement.

This is the true problem with the flawed logic and simplicity of the statement "all accidents are preventable." Depending on when, how, and by whom it is used will completely change the meaning. The context of the statement can take it from a good and honest prevention challenge or it can be seen as a weapon used after an event to establish culpability for the actual bad outcome itself.

In my experience with fatalities, most organizations will carefully craft a statement that puts the ultimate responsibility and accountability for the event solely on the shoulders of the person who died. That action of blaming the victim is not because the organization is mean or trying to escape from their role in the bad accident. Organizations normally have been taught to see the ultimate responsibility for the outcome as a decision made by either the worker or the supervisor of the worker.

In our movie cast study, it is easy to see how preventable the fatality accident would have been after it happened. Therefore some of the accountability for the accident would have to fall on the people who had the accident. This accident was preventable. The problem is the folks doing the work failed to prevent the accident. So, was this accident preventable? The statement is both true and false at the same time. Who said it, when it was said, and about whom the statement was made will dramatically change this meaning of what I think is thought to be a powerful and moral challenge to all the people in our organization to be more preventive and effective in managing their own future.

Is that enough of an argument to persuade you that "agreeing to disagree" is a lazy and dumb answer (I kind of hope the man who said this statement reads this chapter and feels a bit squeamish!) and over-simplifies the context part of the paradox? To me the real problem is if we buy in to this statement as a value-basis for the foundation our safety program, you know what I mean...banners that say "every accident is preventable," hard hat stickers, signs in the change rooms, are we not building a case that the worker is at fault before the worker even has the chance to fail. "If every accident is preventable, why the heck did you not prevent every accident!"

WWLFIWWF to HWFTCTO

Is the statement that "all accidents are preventable" also creating a "chilling effect" around reporting? Does this statement drive our investigative thinking toward looking deeply at the front-end of an events and looking less deeply at the recovery and capacity side of the event? Remember WWLFIWWF!!! (Translation to this statement is What We Look For Is What We Find), do we go in to investigations and corrective actions with a bias towards the belief that the problem is a failure to prevent all accidents? If the answer to this question is "yes, " are we not only looking at only a part of the accident? Finally, if we only look at failure to prevent part of an event, are we not investigating all the prevention possibilities that failed to happen at the expense of not looking at the actual accident conditions that **did** happen?

I think the answer to this question is yes and this yes answer horrifies me in so many ways. I mentioned earlier that I looked at a rather large grouping of fatal accidents during 2016. I was amazed in most every case the accident investigation reports I was given for all of the events all seemed to come to the same conclusions. I actually prefer to call an investigation's conclusion a "judgment of need." I think giving areas that have needs for improvement is better in that these tell the organization what is wrong as opposed to telling the organization what to fix. Either way, all of these investigations had similar findings.

I will make a claim. I have found, as I look at investigations, the same two findings in almost every investigation. Because these findings are almost always present, I have found most investigations recommend the very same corrective actions. If all the investigations have the same findings and corrective actions, we really don't need to keep doing investigations. We can save a lot of time by simply producing a standard investigation, to be used in all cases for all events, and a standard set of corrective actions, to be used by all organizations for all events.

Maybe it is unfair to compare these investigation findings across different organizations and countries. I am not sure we compare fatality and serious event investigation findings anywhere in the world. I certainly had not done many comparison studies on the fatality-side of investigations before. It was striking however to see good investigations, investigations that were performed by professional and caring people with the best intentions, all lead to the same findings.

Every case, every report, every document I saw stated the "root" problem was somehow the worker or workers that were involved in the significant event had failed to identify the hazard and had failed to follow the proper method or procedure to do the work in question.

Let me repeat the lesson from this comment. This lesson is more important than it is novel and if you can learn from this lesson, you won't have to learn from another lesson by actually having an accident.

All of the investigation reports had the same two outcomes:

1. A Failure to identify the hazard.
2. Failure to follow procedure.

This surprising observation of these causal similarities gave me great concern. Here we had the worst possible accident outcome an organization will ever have and all the investigations for these bad accidents could find were two rather uninteresting findings. The worker did not identify the hazard and the worker did not follow procedures. Something did not seem right.

Was it bad investigations? No, that did not seem to be a plausible factor. These were big companies and small companies. All of the investigations seemed to be thoughtful and well intentioned. Was it bad people? No, that also seemed to be too simple of an excuse. The events in question were of all types: falls, confined space, struck by, and the normal litany of bad outcomes. What could it be?

I am convinced the problem is prevention.

I know this sounds crazy, as prevention is what people who are concerned about safety and reliability do for a living, but nonetheless it seems when something bad happens the outcome seems to always be some kind of surprise to the organization. I also cannot say this enough times: Prevention is extremely important and has high operational value. Prevention matters. I am not against prevention efforts. Have I said enough? Maybe, I'll have come back later and reinforce the idea.

The outcome of an event, in my mind and past experience, is almost never surprising. The event may be unexpected, but the actual outcome, the consequence, is predictable and normal. We get all twisted up when we look at an event that tragically we did not prevent the outcome. When two trains collide going in opposite directions on the same track...the accident is surprising but the consequence is normal and remarkably predictable.

WWLFIWWF. The acronym we need for a more enlightened view to fatalities and serious events is HWFTCTO (How We Failed To Control The Outcome).

We didn't fail to prevent the serous accident although this fact is most likely what we investigated and for which we produced correction actions. We failed to control the consequence of the serous accident and probably did not investigate or produce corrective actions to correct this deficiency. We really do find what we are looking for every time.

You are beginning to see the difference between these two ways of seeing the world of workplace safety and reliability. We prevent accidents. We manage consequence. These two statements are a quite different way to think about the capacity existing in our organizations to do high-consequence work, safely and reliably.

Chapter 7

What our Discussion Does (so far...) for Fatalities and Serious Event Thinking. A summary of the first half.

Lets recap some of the thoughts presented in the first part of this book. We have covered a lot of ground and moved quickly around some important ideas, in a sense barely scratching the surface of the analysis and discussion this topic truly deserves. This is a potentially controversial conversation to have around such a serious topic. After all, the outcome of these events is the worst possible news for your organization, for the families of the workers, and for the community in which your organization lives.

We trudged through some ideas that are scary sounding when discussed frankly and honestly, and the reason we trudged so far and so fast is because all we can hope to accomplish with this discussion is the advent of some new thinking about this problem. We want to better frame this problem so we can better think of ways to manage the margins around this problem.

We have to be better at understanding this problem. We must change the way we talk about this problem. Most importantly we must understand the way we ask the questions about fatalities and serious events will ultimately direct and color how we craft our path forward.

We have introduced several ideas that are important to moving our discussion about fatalities and serious events towards a new way of thinking. So far we have built a primary case that, in essence, is helping us to understand where current thinking ends and new thinking must begin.

These are the important parts of our discussion so far:

- *Fatalities and serious events are not normal safety accidents.*

Doing the same things you do to make safety happen – better or harder - will not make you fatality-free and forever safe. If we think of fatalities as the "ultimate accident," then our thinking naturally goes towards intervening earlier in order to prevent the accident from happening. If fatalities where the ultimate endpoint to the safety pyramid, then our current solutions would be suitable and effective and we would never have a fatality or serious event. We know that is not true.

Because we have a tendency to think of these as more *final events* (serious outcome events) as the sum of many additive *transitional events* (small outcome events), we naturally think of a fatality or serious event as an overall failure of the organization's safety program. In reality, these events are a function of system that has somehow lost (or never had) the capacity to fail successfully.

Seeing fatalities and serous events as something other than super-powered accidents allows for new and different thinking and analysis. Seeing these types of events also allows for different types of safeguards and controls. Mostly, seeing these as places where normal operational practice failed to provide recovery allows the event learning to move beyond the failure to prevent the event.

- *Fatalities and catastrophic failures are outliers, anomalies.*

If fatalities and serious events are **not** the logical next step for a linear accident chain, then what are these types of serious events and what is the prediction system we should use? These events are anomalies or outliers. They are not predictive, because the systems in which these events happen are normally stable and safe (enough) to be productive and not deadly. These types of events are actually produced by the very system that does not want to produce these outcomes.

As much as we want to make outlier events predictable, these events simply are beyond prediction. Mathematics teaches us that outliers are a non-predictive outcome. The difficulty is these outlier events to our processes and systems will always look predictable after they happen. Everything becomes obvious after it happens.

- *Anomalies are not predictable – no safety pyramid can ever predict an anomaly.*

As desperately as our organizations desire stability and predictability, no prevention tool has the ability or sufficiency to predict all serious events or fatalities. Our need to have this type of predictive power has, over time, become almost mythological in the belief that we must have missed something in order for such a bad thing to happen. We believe causes that are not true.

That does not stop the retrospective bias that says a bad outcome happened because there must have been some bad choices or actions in the system. This cannot be true. Every time your organization has a failure your workers will have exhibited bad choices and actions. The problem is that every time your organization has successful production, your organization will also have exhibited bad choices and actions.

Think of anomalies as possible but not probable. Operational anomalies are always present in your organization. You will not predict or prevent anomalies or outliers away from the work you do. At best you can only manage the consequence of the outlier when it appears and, eventually learn as much as possible about the outlier and your margins for recovery.

Remember, a **statistical anomaly** is the degree to which something (an event, a person, a data point) falls out of a normal scope for one group - but at the same time the anomaly is a result of being in that same group.

- *The same system that normally creates success has the ability to create horrible failures.*

Our systems and organizations mostly don't kill people. Most of our workers will work their entire career and not have a catastrophic event. The systems your workers interface with on a daily basis don't normally create serious safety and reliability outcomes for the workforce. The idea that this normally stable system has the ability to create significant consequence becomes less and less realistic the longer the workers work without event.

The paradox is that the system that is safe is also simultaneously dangerous and workers, who almost never get hurt, are constantly negotiating risk perception and consequence levels and assessments. Workers are creating safety as they perform work. Safety is not built in to your processes because your processes are paper documents, not work. The belief that somehow workers consciously choose to fail, colors most of our thinking about accidents and is especially amplified for serious consequence events.

When faced with a failure first ask how does this work get performed when it is performed successfully in order to understand the complexities of the failure that has happened. The system was thought to be safe and stable...how did it perform differently from the expected outcome?

- *The strong bias for event prevention is forcing us to always try to fix our "failure to prevent" problem.*

The trouble with catastrophic events is that once the accident has happened we are hard-wired to search out the most obvious cause for the event and call that action analysis. A Black Swan event is easy to simplify after it happens, but remarkably hard to notice before it happens. Our organizations want to look with the power of hindsight and determine where "we went wrong?" That seemingly innocent way of thinking creates big problems for your organization's future.

It is almost impossible for a manager to not struggle with the question,

"Why did we not stop this before it happened?" This question seems really important, but the question is a giant waste of breath. The event has happened – so every answer you gather around the fact that your organization failed to prevent the accident is based on "fantasy" not on what actually happened. Our investigations, our corrective actions, and most seriously our thinking about serious event recovery is dripping with fantasy statements about prevention. That type of thinking has not been, nor will not be helpful to solving this complex problem.

- *When uncertainty is the outcome, manage the parts of the process that are certain.*

You cannot prevent an unexpected event. As reliability and safety professionals, we are constantly struggling to know what we don't know. We are asked to identify and mitigate all the potential conditions and context that might lead to failures, both small and large.

It is impossible to manage the uncertainties that are a part of our organization's production and operations fabric. We cannot possibly be expected to predict all the things that can go wrong. In fact, Plato's notion of uncertainty may most clearly present the true depth of this challenge, "more bad things can happen then will happen." Almost anything can fail, but actual failure is rare and unpredictable.

Since we cannot solve Plato's paradox in the safety office, perhaps we should think about how we respond to this problem. Let's consider the following example: The airlines cannot control the weather. The airlines can, however, manage their response to the weather. The weather is a condition that is beyond prevention and control. We cannot control the weather. The weather in this example is uncertain. The airlines response to the weather is controllable. The airline can have snowplows and de-icing trucks available at airports most likely to have snowstorms. The response mechanisms are certain. We cannot manage the uncertain, so we must manage the certain.

- *You cannot prevent your organization from experiencing a fatal or serious event.*

It is difficult on first pass to make this point, but what urgently needs to be said is this message: We will never prevent a fatality from happening and yet we are constantly preventing fatalities from happening. Confusing, I agree, but almost vital to talk about within our organizations. Prevention is important, but not sufficient to prevent serous events and fatalities. When we prevent a serous accident, prevention is enough. When we miss the prevention potential due to planning shortfalls, hidden hazards, operational drift, or unforeseen events, then prevention does not work. Most events that have serious consequences where not planned to be prevented.

This statement, of all the statements made so far in this discussion, is the most inflammatory. I showed some of this new fatality and serious event thinking to a group of hot-dog safety people and they all said, "that is NOT true – we prevent fatalities all the time!!" and then summarily discredited almost everything else I said. I know this is an important argument; I am just not convinced I have captured the right way to present a new way of thinking about this problem.

That leads to the second part of this discussion. What do we do with this information? How do we make the world a place where people are less likely to die while doing their work?

Part Two

The Great Shift in Thinking about Fatalities

How is it that in good organizations people still get killed? How could it be that we have managed our legal and compliance reporting numbers, days away from work and total reportable cases, to levels so low that they were thought to be unreachable and still have not impacted the events that really matter, the events that kill people? It makes no sense. If our thinking about the low-level events is correct, why does the same thinking not apply to high-level events?

Hmmm..."if our thinking about low-level events is correct why does this same thinking not apply to high-level events..." is a fairly important question to carry forward in our discussion. Are we vain enough to assume, quite wrongly it would seem, that our thinking has been right all along? That would be stupidly narrow-minded of us as people who are constantly questing and trying to learn and improve safe and reliable operations.

> **Why doesn't the safety we normally use to make our organization safer also work to manage more serious accidents?**

Why doesn't the safety we normally use to make our organizations safer - work to manage more serious accidents?

Wow, that obvious question sums up the current problem for safety and reliability professionals all over the world. That is a strong research question.

Why doesn't the same motion that helps to reduce ankle sprains also stop decapitations? The wording and thinking used to frame this problem statement is starting to be more useful to safety professionals. We cannot fix a problem that we cannot identify.

This research question is not new or unasked. These exact questions have been around a long while. We have all sat in those meetings where smart people ponder our future and somebody makes an impassioned plea for us to realize that "safety is about people" or some other moral statement that is meant to drive home the idea that we must not care enough to not kill workers. The problem is when we make this argument personal, moral if you will, we tend to place the emotional need to be better stewards of our people above the practical need to design systems that can fail safely. We don't see a worker who falls from heights as a moral failing, we see the fall as a function of gravity – no moral judgment at all – and therefore we can make a guilt-free case for buying fall protection gear, or confined space gear, or seat belts - you get the point by now I am sure.

This problem, fatalities and serious events, has nothing to do with caring or trying. Fatalities and serious accidents are not a moral failure of the organization. These events are random, non-normal, outliers to what is normally a stable and productive system. Caring and trying are not solutions for this problem, caring and trying is never a solution for this type of problem. Caring and trying are ways to reinforce the strongly held belief a worker was killed or injured because somebody (often including the worker who was injured) did something bad. Here is what we know; you don't have to do something bad to die at work. Workplace fatalities are not moral judgments on the worker or the organization. Workplace fatalities and serious events are operational anomalies with high consequence and insufficient controls.

I haven't found an organization that kills workers because the organization is morally weak. The types of organizations that have serious events are just like every other organization that does not have an anomalous outcome, a bad outcome. When I work with an organization that has had a fatality, I often see a very responsible and concerned organization that has been horribly damaged by an event. The organizations that have fatalities and serious events usually look like other organizations that do stable and predictable work, every working day.

We must look beyond the old view of safety: The problem is deficient workers. We must look towards the new view of safety: The problem normally exists because the capacity to do the work in question was expanded past the point of reliability. Workers got too close to the margin and when the failure happened the absence of margin left the workers no room to recover.

We know our traditional view of fatality and serious event prevention has not been effective. We also know that a renewed emphasis on doing the old protections and preventions more aggressively is not effective. We are then left to discover new ways to make catastrophic events come into some kind of control, even though we can never measure how effective we are. You can't measure, in either new or old thinking, what does not happen. All we want is to not have events happen.

Those questions, however, have not been answered using old view thinking and qualitative observations of high-consequence events like fatalities and serious events. Could it be possible that we as a profession are so entrenched in our old thinking that we could not apply new thoughts to these serious problems?

New thinking about fatalities and serious events is not going to be easy. We owe it to ourselves to really explore where the old thinking comes from and how we got to where we are around these types of catastrophic failures. I have attempted to build a case that because of the need to have ongoing and sustainable safety and reliability perfection, our practices and thinking about safe and reliable systems have moved to an almost complete reliance on prevention of accidents. This preference for prevention, although effective to a point, is not sufficient. Prevention is important, but prevention is not prediction. Prevention is not enough.

Those unanswered questions offer us a new opportunity, a chance to apply new thoughts and ideas to old problems and potentially get different outcomes. This is the reason we should have this discussion. We have a chance to do safety differently and to potentially get much different outcomes. The only problem is we will need to think much differently about this problem.

We will have to start with a bold statement: We don't have a failure to prevent fatalities and serious events. We have a failure to control the outcomes of events that lead to fatalities and serious events. We don't prevent fatalities. We control outcomes of accidents. We must learn to fail safely.

Fall Protection

The entire discussion so far in this book could have been greatly shortened by this one example. Let's start with a basic question; it is an easy question, so jump right in to this discussion. Better yet, ask some folks in your organization this question and see what they say. I think you will be pleasantly surprised by the answer and the ensuing discussion this answer creates.

"Why do we have people wear fall protection gear when they work at heights?"

Lots of folks will answer, "So they won't fall."

That answer is wrong, a bit predictable, but not correct by a country mile. Fall protection gear is not designed and used to help the prevention of a fall. Fall protection gear has nothing to do with the prevention of a fall. Fall protection gear actually manages the speed and the force by which the worker who has fallen hits the ground (or the end of his lanyard). Fall protection gear controls how workers land.

Our fall protection programs are more forward thinking than our fatality and serous injury protection programs. We see falls as unpredictable in the specific details but probable in normal operations. A fall is truly an outlier or anomalous event. We can't know enough about the work to know who will fall or when the fall will occur. Because we don't have that information, we don't manage hazard prevention; we manage consequence of the event.

We don't manage fall protection by asking workers to not fall. Asking workers not to fall is perhaps the least effective safeguard we can put in to this safety program. If asking workers not to fall were effective, that would mean falling is some type of conscious choice the worker makes right before they fall. I know as you read this, this statement seems incredibly juvenile; however, when you really deconstruct the idea of asking workers to be more careful when working at heights, being careful is not dependable.

We actually assume the worker will fall and manage the consequence of the fall itself. Let me state that again, we design the safety protocols around the assumption there is a 100% chance the worker will fall, and when he or she falls we will have a harness available to catch them before they die. We don't prevent falls. We can't really prevent falls.

Fall prevention is actually ineffective in preventing falls. Fall prevention is effective in preventing uncontrolled and dangerous worker landings. Once people started to figure this out--you don't really prevent falls, you actually expect falls to happen--then those people quickly begin to understand that you must manage this hazard interface by controlling the event outcome.

In other words, we have created a system that has the ability to "fail safely," or should I say, "Fall safely."

A False Sense of Security

Earlier we talked about prevention bias, the belief that when a bad thing happens there is clearly some failure to prevent the bad thing from happening. This thinking is harmful because it gives good organizations the false sense their systems are well designed and stable--the false sense that if workers would do as they are told nothing bad will happen to the workers or the organization. Prevention effort is like that; prevention wants you to think that you have a duty to act on the front-end of the work. The better you get at prevention the less you value or need recovery efforts. The better you get at prevention the stronger your belief becomes that your organization knows they can prevent events. Remember we live in a world where we tell our leaders and workers all accidents are preventable.

This idea that if we plan better we will prevent better is not a bad idea. In fact, there is much good to be gained from the power of preparation, planning, and prevention. The problem is that plans and prevention activities cannot plan and prevent unexpected events. The idea that unexpected events could be planned and prevented goes against the logic of unexpected events. If you expect the event to happen, you plan for the expected – always with the nagging realization that a complete prevention program and work planning action is impossible.

Suddenly, your organization is faced with a permanent dilemma. How can you do the impossible and do it well and within your budget? The answer is that you will never completely be able to prevent every possible operational conflict out of your system. You have known that you could not be perfect in your prevention for a long time; this is no big news for you or your leaders and workers.

The solution has traditionally been to rely on the most effective (and in my observation probably the most operationally conservative) problem-solving tool you have in your organization: your workers. These workers are relied upon to manage all the operational anomalies and variability in most every workplace that performs some type of work. Workers are adaptive. Workers are mission propelled. Workers are normally excellent problem solvers. Workers make weak, snagged, or broken processes and plans function, and they do this magic all the time.

Workers add to the false sense of security by simply being normal, adaptive problem solvers while doing work. Workers are so good at adapting to, and monitoring for, hazards in real time. We can never underestimate how vital a smart worker identifying a potential fatality or serious event is in de-escalation of potential failure. What workers do is keep failure from succeeding. Knowing a great amount of your prevention effort is directly tied to workers actively preventing events is a great reassurance to leaders all over the world.

The problem is workers are not as reliable as a machine guard or a fall-protection harness. Whether we want to or not, we have attached fatality and serous event prevention work to the most adaptive and least reliable part of our organization. It seems unfair to spend time asking why we failed to prevent an event when the workforce in your organization is always engaged in event prevention. It is not why we failed, but better phrased as how we missed this important piece of information with so many prevention experts doing the work?

It is easy to say we are good at prevention. The reason you can say your organization is good at prevention is because your organization is really good at prevention. Your good at prevention until you are not good at prevention, and then you are bad at prevention.

This tension between prevention and recovery all started to become a bit more interesting to me when I went to the field and looked at examples of prevention and planning in action, at real workplaces with real workers.

Prelude to Three Capacities

I mentioned earlier in our discussion that I work with companies who have had catastrophic events. I don't really help them investigate. I don't really tell them what to do in order to fix all the problems that they think they have. What I do is help restore this organization to the ability to do high-risk operations better then they did operations before the accident. I help organizations think differently and ask better questions as a part of the restoration process.

As I did couple of these types of peer assist visits, the word must have spread around because I started to get calls from other organizations that wanted me to be a part of their process of restoration after some serious consequence event. Helping to think about the recovery differently really helps the organization change the way they think about what they need to do next in order to restore and recover their operations. I am convinced organizations only really manage two things: The organization's confidence to believe they can do this work again and the organization's capacity to fail safely.

In looking at many fatality events in which I was privileged to be invited to help the organization begin the restoration process, I started to see some important themes appearing in each of the event description, interviews, and investigations. It became clear to me none of the events I was working on happened because the organization had bad planning or prevention habits. In fact, for about half of the events the planning and prevention program was the best I have ever seen in my entire career, truly world-class safety and reliability planning and prevention.

These same themes would arise over and over. The accidents were different. The companies and the products the companies made were different. The locations and cultures of the companies were different. Yet, there seemed to be these themes that appeared over and over again.

It soon became clear to me that of all the events I was seeing, and it was a lot of events – the most I had ever seen in a time span of about 60 days – there were really three very distinct ways that fatalities and serious events seem to happen. These three reasons why people die or get seriously injured at work are not very technical nor complicated. In fact, these three reasons seem quite normal.

To start our discussion let's begin with three quick case studies to be used as illustrations. Again, don't look for how the organization failed to prevent these failures, but instead look for a theme around the organization's ability to extend its margins of recovery, its ability to fail safely. These are quick but should illustrate the point quite effectively.

Illustration One – The Falling Man

Cleaning tanks in a refinery is a big job. Tank cleaning is the kind of work that is never completed. When the tank cleaning crew finishes one tank their reward is to move quickly to the next tank to be cleaned. There is never a shortage of tanks and that means there is never a shortage of work.

The process of cleaning tanks is risky, dirty, and requires rigorous protective equipment for the workers who will enter the tank to clean it for its next use in product production and manufacturing. It is hot, uncomfortable, and stuffy. The workers work long shifts and look forward to the only break in their day, lunch hour. The rest of the day is spent cleaning tanks.

The tank cleaning process is a normal, almost daily operation and happens continuously in one of the hundreds of tanks at this refinery. All the tanks look similar and because of the material this refinery was using, the tank markings were often difficult to see, and at best case scenario, the markings where only visible for a couple weeks. The tank crew also was used to re-sign the tank with stencil and paint.

This process is so common that at this specific refinery, the tank cleaning work had been contracted out to a company whose only job was to constantly be cleaning tanks. The tanks were basically opened and then scaffolding was built inside the tank so the cleaning crew could have the safest possible environment in which to do their power washing and brushing.

The cleaning workers entered the tanks from the top after the scaffolding crew had completed the building of the scaffolding and had done and signed off the inspection for safety. The top entrance was necessary because the workers then worked from the top down to clean the tank, which also allowed the excess water and material to drain as the crew moved down the tank. Because the top hatch was above the scaffold and because there needed to be room for the cleaners to stand and clean the top of the tank, the scaffold was actually placed about one meter below the hatch. The practice was to back through the hatchway; legs first, and then drop to the top-level of the scaffolding. It was not a far drop, only about 30 inches for a normal sized man to drop on to the scaffold and begin work. Total time to clean the tank was a half-day. A good crew was expected to do two tanks a day, every day.

On the day of the accident, the two-person crew was on their second tank of the day. The crew had enjoyed a relaxing and cooling lunch in the lunchroom and was doing their pre-job briefing and preparation to start what would be the second tank of the day. The tank number was on the work order and the pre-job briefing document; the tagging for the scaffolding was identified and confirmed. The crew climbed the ladder to the top of the tank platform and identified the tank. The crew knew that the next tank to be cleaned was one tank to the *left* of the tank they had cleaned and finished that very morning. The crew leader reminded his coworker to be "100% hooked up" before the two workers donned their respirators and face shields.

The workers then identified the tank they had completed that morning and then moved one tank to the *right* and removed the hatchway cover. When the cover was removed the lead worker then placed his legs in the hatchway and dropped to what he thought would be the top of the scaffolding.

Sadly, the tank the crew-leader entered was not the correct tank; rather it was the tank on the other side of that morning's tank that was the correct tank. The worker then fell some 35 feet to his death. Rescue attempts were commenced immediately but the worker did not survive the fall from the hatchway at the top of the tank.

Illustration Two – The 100-Dollar Router

It was a normal Wednesday afternoon at the computer support center for a major, low-cost carrier in the United States. This airline was famous for its on-time record, fun-loving employees, and incredible dependability. A Wednesday afternoon was seemingly normal for this airline. In fact, Wednesday was not even a big travel time. If anything the airline was not using all its capacity and was getting some needed breathing spaces for not just their people, but also their computational systems.

It was a normal Wednesday until it wasn't.

When high-performance computing systems are designed they are designed with a robust recovery strategy. In most cases, as was the case with this airline, there is a complete duplicate and sometime triplicate data center located in a different place with a different power source and a robust switching process that allows for a seamless jump to another system if there is a problem detected. This process is so effective and mature in most cases, the consumers of the system would not have known that there was some type of problem and that a dramatic corrective action had taken place. It happens automatically and instantaneously. It really is quite impressive. This airline prided itself in having one of the best and most robust systems available in the industry

These secondary systems are tested and retested. As a matter of course, the computing people actually put problems into their systems on purpose just to see how well these systems recover. There are few industries better at highly reliable performance then computer data centers.

That is until Wednesday afternoon. On the Wednesday afternoon in question a small router, a router that could not have cost much more then a hundred bucks, almost failed completely.

I say almost failed completely because had it failed completely, this story would not be told. If this router had actually fully failed, the secondary system would have been alerted and taken the functionality for the entire computer system and moved it to the secondary system. Had it failed the router would have been identified and replaced quickly. An after-action review would have been done on the switch-over to the secondary systems, notes would have been taken, and eventually the functionality would be placed back on the primary data center, had it failed completely.

2,300 flights were canceled and as many as 300,000 people where stranded because a router stopped sending data but kept sending the signal it was on-line and functioning properly. The only example I can even think to describe this is like this: have you had Wi-Fi signal but not Internet service? Your phone thinks it is connected but in fact it is connected to nothing but the Wi-Fi signal. This is what happened to this airline on the Wednesday afternoon in question. The signal was connected but there was not data being delivered. Had it caught fire or blown-up, the system would have caught the problem, but partial failures on routers are rare, they tend to fail bi-modally – routers either work or routers don't work - and so no detection strategy had been placed for this specific failure mode.

Illustration Three – The Falling Door

"It was a brand new aircraft maintenance hanger. It was only 5 years old and it was never used for the actual reason it was built. "The company got it for a song because most people don't need a building with two 95 foot doors on the front of the building." This was what the building manager told the investigators as they began an investigation for a fatality in which one of the 95 foot doors fell off its hinges and landed on some workers, seriously injuring two and killing one.

The door had no history of problems and had functioned perfectly every time it was used. Since the building was relatively new there were no maintenance activities on the door and none scheduled. It was as if the door just fell off one day with no warning or reason. Gravity had overtaken the door's engineering and the door fell.

No problems with the other door were identified. No problems with the door that fell were identified. The door simply fell off its hinges and landed on the ground on top of three workers. There would be a big engineering investigation. The door manufacturer was involved. The building owners were involved. The engineering firm and construction company were also involved in this investigation.

After the eventual, deep-dive engineering investigation process was complete, the findings seemed to indicate a failure in the manufacturing of one of the hinges used to hang the door on its track. The other hinges were all tested and replaced. No faults where discovered in the existing and functioning hinges. The hinge replacement was completed for extra caution.

The doors were reinforced and additional safety strapping was connected from the doorway to the actual doors to prevent a free-fall accident from reoccurring.

The Three Reasons We Have Fatalities and Serous Events

There are three "buckets" that kept emerging and re-emerging in every review of these serious events. It only becomes apparent when you start to trend many events beside each other that the problem is not a "failure to prevent" the accident. The problem is in the way the systems and organizations recovered from the actual failure.

The three illustration case studies each represent one of the three reasons we have catastrophes. Each of the three illustrations is an example of a serious failure event, actually a serious failure with catastrophic consequence, yet each of the three events seems different in the way the problems play out. One is a clear conduct of work problem, one is an engineering failure, and one is almost an act of God; yet, as a triad these illustrations represent how fatalities and serious events happen in organizations.

Each of the three illustrations have a strong component of "not preventing" the events to be sure. The failure to prevent is probably important to note as is the fact that the organizations failed to predict this particular failure modality and therefore were caught by surprise. My question is how important is the identification of what we failed to prevent after the accident has happened.

What is more interesting, at least for the discussion we are having, was the complete lack of recovery – even when the system was designed to recover. The ability for the event to happen is surely a direct function not of a lack of prevention, but more of a lack of controls...or margins...or safeguards.

Here is what I found when I started to really dig into the idea of consequence recovery as a strategy for reducing fatalities and serous events. These are the three categories of recovery failure. Seemingly every event I have observed fits in to one of these three definitions:

1. Absence of Controls
2. Insufficient Controls
3. Event Beyond Safety Basis

None of these categories have anything to do with prevention of the event and everything to do with the ability to recover, or as the great David Woods calls it, "graceful extensibility." This started some new thoughts. What would happen if we simply stopped focusing on prevention and started focusing on recovery? Would we get worse? People are already being killed; it can't get much worse than that, can it?

I started to really dig in to these ideas. Perhaps, this story is not complete and there will be more categories, but every event I observed had one of these three characteristics present in the story of the organization and their event. Seeing an event from the safeguard side does allow you to think much differently about catastrophic failures and how they are so unfortunately successful. The different viewpoint also moves the findings and corrective actions to a different level as well.

The three illustration case studies were each selected because they are examples of each of the three event reasons. Allow me to identify the category and why I think the illustration is an important example"

1. The first illustration, The Falling Man, is an example of **absence of controls.**

    No safeguards or controls are present in this event. As you read this case study I am sure you noted the myriad of complexities that exist in doing this type of dirty and pressurized work. The question must be how could we put a worker into both working at heights and confined space without any of the normal, even basic, prevention or planning and preparation that this type of work hazard would require? The answer is in the inability to identify the tanks, to be certain, and human error, going to the right instead of going to the left. These factors are not hidden or unimaginable to the worker or the work planners. Had the crew gone to the correct tank this event would not have been an event. It was as simple as the difference between right and left.

    The context of the event was brittle; it would not take much to trigger this type of failure. I like the use of the word brittle in that it implies there is little flexibility in the context of this event. Any type of variability could be enough to create an environment where you mistake left for right.

    The problem in the context was the complete lack of controls, perhaps not for the fall but for the identification of the correct working tank. Planning to fail allows a certain amount of flex to be created as capacity within the work systems.

2. The second illustration, The 100-Dollar Router, is an example of **insufficient controls.**

This illustration is a lesson in robustness of controls and effectiveness of safeguards. The idea of a false sense of security comes up over and over again as you review fatalities and serious events. This could be called the same false sense "complacency" on the part of the organization or even the specific workers doing the work. Either way, the learning for us is that the controls must be robust enough to actually provide the recovery should the recovery be needed.

Commercial planes have lots of thrust that is produced by their engines, so much thrust that if an engine goes out there is enough capacity to still fly the plane. Planes hardly ever need this extra capacity — until they do need this extra capacity. There is built in extensibility, graceful extensibility, built into a commercial airliner. This example is supportive of sufficient controls present when the work fails so as to allow the work to fail safely.

3. The third illustration, The Falling Door, is an example of an **event beyond safety basis**.

    No person could have imagined that event was even possible until after the event happened. The event was beyond the imagination of the people who are charged with imagining ways events can happen. In my world, we would call this type of event an event beyond our safety basis documentation. This event is a function of a failure that simply could not be prevented because it could not be conceived.

    For events that are simply beyond the ability for the organization to imagine, there is no prediction and therefore there can be no prevention. The organization's entire system is dependent upon the organization's ability to recover from the event safely and quickly. Don't underestimate this category; it would be easy to see this final bucket as a "catch-all." I am fairly sure the category is not an also ran, but in fact a fairly important learning for the safety and reliability world.

Our ability to build robust recovery is not based upon what may happen. Everything may happen or nothing may happen. Our recovery is more about asking if the current controls are enough to keep you from getting seriously hurt or killed.

These three categories of potential failure for fatalities and serious events are by no means predictive or to be used as an assessment standard. What these three categories help us really do is not much more then help emphasize the importance of understanding the controls that either were not present, were not robust, or where not imagined. As I continued to observe other events, these three distinct areas of control have arisen over and over again.

These categories have assisted organizations that are using these new ideas to better understand how much emphasis and direction to direct towards managing controls and safeguards as a strategy for restoration after a significant event.

What work matters? Should I just control everything?

This is an important question for any manager or worker to ask. You can't manage all risk. You don't need to control everything, but to have this conversation we must first try to identify what tasks should be most worrisome. No organization has the ability (or the resources and finances) to fix all its problems. I am not even sure fixing everything is even important to try to do.

So we are left with our normal risk-ranking activity. A good organization probably has created some type of equation to figure risk – in a traditional western-economic model we will try to identify the probability of something bad happening multiplied by the potential consequence of an event. The process is linear, predictable, neat, and tidy…but we are left to wonder how much this risk-ranking idea is dependent upon the prevention bias.

I think our traditional risk matrix model is not as effective on the management of safeguards side of accident or event thinking. Risk ranking is somewhat effective for prevention and planning, but risk ranking is almost always done to manage resource use and not really operational exposure. Risk ranking is generally silent on the ideas of controls or extensibility and recovery. Risk ranking tends to look at the front-end of an event.

Knowing what Work is Dangerous Work is Important, but not like you think...

Dangerous work you can safeguard or control is not really dangerous work. It may seem to be dangerous, but if the controls are in place the activity does not have the potential to actually cause harm. The tasks the workers perform may have significant hazards. It might look super scary and super crazy. If you can control for a negative outcome, make it so the work has the ultimate ability to fail safely, the work would have to be seen as less dangerous, ultimately.

> **The things that kill workers are not the things with the highest perceived risk....**
>
> **The things that kill workers seem to be the things that are the most difficult to control.**

Dangerous work you cannot safeguard or control is really dangerous, and worse even still (if that is possible), completely beyond prediction and assessment. The problem is not lack of prevention; the problem is lack of control. We manage the energy of the event...we always manage the energy of the event.

This way of assessing risk has really changed the way I see the world. I used to think high-risk work was high-risk work. I now know high-risk work with robust controls is not really risky. Should I repeat that statement? *High-risk work with robust controls is not risky.* This statement should be an indicator of the dramatic shift that is happening with fatalities and serious event understanding. If the risk outcome is controllable, the risk outcome is then in our control. If the risk outcome is difficult to control – that risk is scary and non-recoverable.

Where once I thought high-voltage electrical work was extremely risky, I now have changed my thinking a bit. I know that high-voltage electrical work carries risk, but the risk is well controlled with processes like LOTO, arc suits, PPE, grounding, and the like. This work is understood as work that could fail; therefore, the controls around this work manage, not the probability of the failure, but the inevitability of the failure.

Safeguards don't manage the probability of failure; safeguards manage the inevitability of the failure. This is no small statement. We don't manage uncertain, high-consequence work based upon probability of failure. We manage uncertain, high-consequence work based upon the idea that the process or task will fail, the inevitability of a failure.

TThis thinking is flatly different from the way we have traditionally assessed risk. Hazards that are controlled are not as risky as hazards that are not controllable. Suddenly, the world starts to shift towards prevention of events as our priority, to the control of consequence as our new delineator.

Suddenly, activities like worker on foot become much more high risk using this new model to assess both the hazard and the risk. Worker on foot, or facility pedestrian traffic, is hard to provide effective and stable safeguards around. Worker on foot does not easily translate into the graceful extensibility of a control when failure happens. I am not sure what controls we can rely on for a worker versus fork truck accident. There is no LOTO or PPE for getting struck by a moving object. This hazard is much, much harder to control. Because this hazard is more difficult to place safeguards in the normal working context, the hazard level must increase until we can find a way to manage away the consequence of worker contacting a piece of industrial equipment.

By way of illustration for this worker on foot problem, a rather large facility with many industrial vehicles and scads of workers on foot was struggling with a way to ensure that workers would not get struck and injured by mobile equipment in the plant. This facility worked diligently for years writing rules around walking and driving. All the rules counted on higher levels of attention and caution by both equipment operators and plant pedestrians – none of those rules were robust. The solution was simply to ask all personnel to be more alert. By now you know that alertness is not a control. Using this new thinking, the thinking that moves from the prevention bias towards the recovery bias caused this facility to put "cow catchers"[1] on the back of their mobile equipment that actually moved pedestrian out of harms way when the workers on foot got too close to the equipment.

That difference seems really important for our organizations ability to have an effect on fatalities and serious events. This difference represents a huge shift in thinking around the relationship between hazard assessment and resource management. The basic, more fundamental, questions are changing the way we assess risk.

Where once we would figure the potential severity of an event, we are now calculating and assessing the amount of safeguards that are present when the work fails. That difference is probably worth your reading this discussion. Where once we would have had a chart that called an operation high, medium, or low risk, we now will have a chart that helps identify the controls and safeguards needed to do this operation.

It is all about the non-recoverability

Soon, in our discussion we are going to discuss some ways to assess risk and manage defenses and safeguards in order to make work, not really safer, but actually make work have the capacity to safely fail. This is a

---

[1] In railroading, the **pilot** (also known as a **cowcatcher** or **cattle catcher**) is the device mounted at the front of a

different way to think about risk and hazard--at least different from the traditional safety approach that industry has used for years.

We do have a bias towards prevention, just as we have a bias towards hazard mitigation and removal. The belief in prevention and removal is not bad; this belief is simply a little-bit one sided. We need more and so we went out to find more.

Our traditional hazard identification programs tend to find "energies" in our facilities that could cause harm to people or production. Once we identified what it was that could hurt us, we then would start to remedy that hazard away from the work and the workers. If we could make the hazard go away, that was a perfect solution – no hazard – no harm. If we could not make the hazard go away, we created a hierarchy of ways to reduce the likelihood of some type of bad outcome. It was (and is) a seductively attractive system.

With one unusual limitation--our traditional hazard identification programs tend to be rather linear and mathematical. We look at hazard management as an engineering function, and surely hazard management is both greatly served and greatly owned by engineering, which discourages us from seeing hazard management as an adaptive function of workers doing work.

When you look at hazards adaptively, the fluid nature of harm moving in and out of work while work is being performed, you are challenged not by the probability of the worker "hitting" the hazard, but really by the inevitability of the hazard "hitting" the worker.

This adaptive understanding of hazards adds another dimension to our ability to assess hazard outcomes. This new dimension doesn't care much about distance and opportunity; this new dimension assumes the hazard will eventually lead to a failure. The new dimension thinks about recoverability.

Safety and reliability people don't really care about failure that has no potential consequence. We probably are too busy and too resource restrained to care about low-risk, low-consequence failures. Workers don't care at all and probably don't notice the multitudes of failures happening that don't cause bother, create harm, or matter. Who has the time and energy to care about events without outcomes? Hazard with recoverable and controllable outcomes are not really hazards at all

So, if a hazard with a low-importance, low-consequence outcome is not a hazard, then the same would be true for hazards for which the consequence is recoverable. If we can put the toothpaste back in the tube, did the toothpaste ever really escape? Recoverability allows us to do high-risk work, fail, and recover safely. Recoverability is important to the assessment and management of hazard in our organizations.

Recoverability is the ability to bring your process or operations back from a failure to the condition the processes or operations were in without consequence. Recoverability seems complicated and expensive and it is...but recoverability is important. Talk to the Airline in our illustration case study on insufficient controls, and they will tell you all about recoverability and expense.

Recoverability scares organizations because it sounds expensive and difficult to accomplish. It is difficult to ensure every system has a back up. It is even more expensive to test and maintain those back up systems. This expense only matters if you need stability. Instead of looking for all the places where you do recover, look for the places where you cannot recover.

Non-recoverability, however, should be a word your organization uses every time your organization talks about doing work. Do not go out and look for recoverability; you will become depressed and frightened. Do go out and look for non-recoverability. It is in the places where you are doing non-recoverable work where controls and safeguards are most important.

A non-recoverable event is an action that when triggered causes immediate and certain harm to people or processes and it cannot be stopped. A non-mitigated gas leak from a high-pressure tank is a non-recoverable event. Immediate and certain harm from which there is no return. If a worker drops a hammer and the hammer falls into a scaffolding net this hammer drop event is recoverable. If the same worker drops the same hammer and there is not netting on the scaffolding, the hammer will obey the laws of physics and fall to the earth. A hammer drop without scaffolding netting is a non-recoverable event; once the hammer escapes the control of the worker there is no way to retrieve the falling hammer. Non-recoverable events demand safeguards and controls.

Determining where your organization's systems are recoverable is a good thing to do and you should do that assessment. Knowing where your organization's systems are non-recoverable is vital. You have to know where you are not protected. My general rule is that if the consequence matters, make sure the process is recoverable.

Non-recoverable systems are no more likely to fail then recoverable systems; however, non-recoverable systems are much more likely to fail when they do fail, catastrophically. Fatalities and serious events rarely happen in recoverable systems, not because the systems can lead to fatalities and serious events are more stable, but because these high-consequence systems are often more recoverable.

Chapter 9

Risk Defined and Discussed

Risk and hazard are different words.

Safety people have traditionally used the words as if they were synonymous. Hazard is not a synonym of the word risk and risk is not a synonym of the word hazard. The confusion probably has more to do with the way the insurance industry looks at risk management than safety people being intellectually lazy. We tend to see risk as a product of the environment, which is simply not the case. Risk is different from the hazard yet risk exists because of the hazard. Lets look at one discussion of the word hazard included in the following quote:

> The meaning of the word hazard can be confusing. Often dictionaries do not give specific definitions or combine it with the term "risk." For example, one dictionary defines hazard as "a danger or risk" which helps explain why many people use the terms interchangeably.
>
> There are many definitions for hazard but the more common definition when talking about workplace health and safety is as follows: A **hazard** is any source of potential damage, harm or adverse health effects on something or someone under certain conditions at work. Basically, a hazard can cause harm or adverse effects (to individuals as health effects or to organizations as property or equipment losses).[2]

Hazards can be reduced by controls. Hazards can be controlled by

---

[2] Taken from CCOHS Discussion on Hazard and Risk. I happen to think this discussion is elegant and effective and that is why it is include under this cover.

organizations. Hazards can be managed, removed, and mitigated. The way you make hazards get better or go away is to control the absence or the presence of the hazard.

Risk is reduced by certainty. Risk is never managed by the organization – that is personal, worker exposed risk is managed by the worker. Organizations don't control risk. If hazard is a thing that can hurt a worker then risk is the distance between the worker and hazard. Organizations can't control risk. Risk is in the eye of the beholder. There are lots of books on this topic; you could read for several weeks about risk, suffice it to say that risk and hazard are not the same.

We have heard our whole lives that risk is probability times severity. That is a sweet way to look at this, and better yet calculate risk for an organization that sells insurance. It is not a good way to look at risk for a worker doing work in your organization.

Risk is better defined as the degree to which a worker is facing uncertainty. The degree to which the worker faces uncertainty is really difficult to assess. We always have a difficult time quantifying things that don't want to be counted. We don't know what will or will not happen. Since we can't know uncertainty, perhaps we best assume that the probability of the uncertain outcome actually happening is 100%. We have reduced uncertainty by assessing the outcome as certain which shifts our focus from management of statistical probability to the more modern emphasis on ensuring and management of the actual controls and real safeguards.

Safeguards and controls help reduce outcome uncertainty.

A different way to assess operational risk

> *Question One: When you do your work, what hazards do you encounter that have the potential to seriously injure or kill you?*

Question Two: When (not if – but when) that happens, because our processes are not perfect, our work is not perfect, our tools are not perfect, and humans are prone to making mistakes, when your work goes bad what keeps you from getting injured or killed?

Question Three: Is that enough?

We have not been good at assessing risk in our operations. Historically, we have struggled with all sorts of mathematic and quasi-scientific ways to understand probabilities and harm in our workplaces. Part of the problem has been the confusion between the idea of hazard and risk. Part of the problem has been the strong need to quantify risk so we can better manage limited resources. Part of our problem, to put a final and sharp point on this issue, has been our reliance on prevention activities to solely and completely protect our important workers and assets.

Let's introduce a new way to do risk assessment. This new method is not more technical; in fact, this new method is probably easier and more understandable then our traditional methods of assessing risks. This new risk assessment method relies on a series of questions directed to the workers doing the actual work at risk. Who better to help in assessing the presence or absence of controls then the people who actually do the work? When in doubt, ask the people who do the work. They always know the story of how work is really being done and at what cost operationally and physically is being asked of the workforce.

This process begins by asking the workers to think of the work they do that has the highest potential to seriously harm or kill them. Then ask them what will actually harm them in that specific work. This question is important because you are really accessing the deep knowledge of the work environment. The things that kill workers are often not a part of the formal hazard assessment. Workers will tell you what they believe has the highest potential for catastrophic failure. Listen carefully to how the worker answers this question...there is much data available here free for the taking.

Allow the workers the chance to really think deeply about the hazards they face while doing this work. This discovery is as good as an assessment as any other assessment you could do for this specific work with these specific workers. As the workers begin to identify hazards, listen and capture this information in order to transfer this information to the organization's hazard identification process. Allow the workers to provide a voice for the assessment.

> *Question One: When you do your work, what hazards do you encounter that have the potential to seriously injure or kill you?*

> *Answer One: High voltage exposure! We often have to get close to the high-voltage lines when we perform maintenance in the field – and sometimes there is no way to cut the power.*

As the workers begin to identify potential life-safety hazards, the dialogue normally grows and builds. Workers really have thought about the things that can cause significant harm. Workers know what hazards create fatalities and serious events. Having them assist in the hazard identification, although probably not complete, is a great way to shift the organizations focus away from the planning for prevention bias that has so dominated our thinking for the past 50 years.

Once the workers have talked about the actual hazards they face while performing work and after you have captured and followed-up on any additional hazard identification information needed, you can shift towards the second question. This question will help you assess your current processes, policies, training, and awareness around the identified hazard.

The second question requires a bit of narrative before you address the workers. The second question moves your organizations language from uncertain to certain. Do not ask "if" your system fails, instead ask "when" your system fails. This moves the assessment discussion from imagining the uncertain, potential outcomes to managing the certain and actual outcomes.

It is important here to introduce the idea that when you talk about an uncertain, operational future it is important to use specific language. Do not use potential language to talk about potential harm – the use of potential language simply reinforces the notion that the event probably will not ever happen. Instead use precise language to talk about uncertainty – this allows the worker to imagine the event happening to them. That difference is extremely important to the process of assessing potential fatality and serious event conditions.

On the second question say, "**when** this system fails" and then give the work team the reasons the system will fail. Examples like uncertainty, variability, change, imperfection, process flaws, human nature, and bureaucracy all provide real reasons for work not happening like we thought the work was going to happen. This background conversation and discussion takes the attention away from blame, worker error and human fallibility and allows the workers to actually imagine the failure happening.

> *Question Two: When (not if – but when) that happens, because our processes are not perfect, our work is not perfect, our tools are not perfect, and humans are prone to make mistakes, when your work goes bad what keeps you from getting injured or killed?*
>
> *Answer Two: Well...Lock out and tag out is really important, PPE makes a huge difference, grounding the work before we touch it is vital...verification and clearance is most important as well...insulated tools, training, my experience, my buddy watching my back...being really aware and taking the time to do the work*

safely all keep me from getting killed if the system totally fails.

Wow, that answer is not only informative but also vital to understanding just what exists in our processes and systems that keeps people from getting killed every time the workers perform high potential work near and around hazards. This answer is not ever right or wrong. You are not assessing the worker's knowledge. You are assessing the presence and robustness of your organization's safeguards and controls. Knowing what the workers know that keeps them safe makes it easier and better to manage and assesses your organization's controls for both presence and effectiveness.

The third question is the most important question for the new view for understanding and potentially reducing fatality and serious events in your organization. The third question asks simply, "Are you protected enough?" As simple as this question is, this question is vital to understanding the sufficiency and robustness of the controls that you have in place within your organization.

Question Three: Is that enough?

Answer Three: Yea, for high-voltage work we have a really good, really strong process and it works. It would be bad if we changed our process. I rely on what we do to keep me safe and it works.

This third question, "is that enough?" is an extremely important question. I hesitate to say this, but I am convinced that the sufficiency question may be the most important question that we ask when understanding hazards, assessing risk, and addressing controls and safeguards.

These questions are amazing, not because they are super targeted or scientific, but because they dig a couple levels in to the hazard. What will kill you? That is level one and almost every organization can answer that question. What keeps you from being killed? That is a specific question used to identify the controls for the hazard the workers have identified. Is it enough? That question is the deeply probing investigation into the real tools that help organizations manage fatalities and serous events. Do we have enough capacity to manage the hazard in the actual performance of work?

That question makes all the money. That question, and the subsequent answer is the difference between a good nights sleep and a sleepless night. That is the question that every leader should ask and every worker should answer.

Let's pick another example that might have a bit different outcome. Let's address a hazard that may not have the same level of recovery maturity as high-voltage electricity.

Remember this statement from earlier in our discussion. Now this statement becomes powerful and crystal clear. This is the statement that changed the way I viewed and understood fatalities and serous events. Read it again, carefully. This time with new eyes:

> The hazards that kill workers are not the hazards with the highest perceived risks; the hazards that kill workers are the hazards that are the most difficult to control.

Now, let's look at this second example with the power of the risk and control statement as our guide. This second situation is real and comes from a crew that identified an even more complex hazard for their facility. The amazing thing about this example is that most of us have this same hazard. We all share this whether you have a plant floor or a parking lot...

> Question One: When you do your work, what hazards do you encounter that have the potential to serious injure or kill you?

> *Answer One: Worker on foot! The most risky thing we do is walk through the plant at the same time all the industrial vehicles are moving product. If one of those trucks hits you, you are a goner.*

The workers in this example had just been through an event where a co-worker was killed by a piece of moving equipment. This hazard was fresh on their minds. The prevention approach for this hazard had been to create a "no person zone" around all moving equipment. In this case there was a 10-foot zone around all moving vehicles where the worker was never to enter. The process was rigid, not quite to the level of a cardinal rule, but if a worker was caught within the 10 foot zone there would be soon, certain, and swift consequences.

> *Question Two: When (not if – but when) that happens, because our processes are not perfect, our work is not perfect, our tools are not perfect, and humans are prone to make mistakes, when your work goes bad what keeps you from getting injured or killed?*

> *Answer Two: The driver sees me and I see the driver. The other potential protection we have is the 10-foot safety zone, but that is really weak because the truck is moving and I can't get away fast enough to maintain the 10 feet. Really, all we count on is visual identification. I see the driver and the driver sees me.*

Well, that answer was humbling but honest. The worker's had identified some of the complex realities of the 10-foot rule. The 10-foot rule probably made great sense when it was conceived in a conference room, however it lacks operational fidelity when applied to real work.

The assumption that a worker would enter the zone on purpose to get struck by a piece of industrial equipment is flawed and foolish. Asking a worker to not enter the 10 foot zone assumes that the worker who did enter the 10 foot zone was choosing to violate the worker of foot rule. Clearly, entering the 10-foot zone was not an intentional violation. Entering the 10-foot zone was a fact of having to do work where industrial vehicles also must drive in order to do the organization's work. There is very little worker OR driver choice involved in this prevention. However, the 10-foot rule and this rule's awesome power almost was the entire prevention strategy for stopping these people from getting hit by industrial equipment. If I were to sum this up in to a couple of words, I would say management asked the workers to care more and move faster, or is it move faster and care more.

That set the discussion up for the third question. I was really interested in what this response to this third question would be.

Question Three: Is that enough?

Answer Three: You kidding? What worries me is the day when I see the driver and I think the driver sees me. We both look, but one of us does not see. That is the day that I worry about. The driver will not hit me by asking me or the driver to be more attentive...nobody knows when he or she should be more attentive until after they need to be more attentive.

That was the day this plant explored two additional and more robust controls: Rear collision avoidance systems and nerf-covered rebar that pushed the workers out of the way. Eventually, the plant purchased RFID implanted plant ID cards that alerted the industrial machinery operators when a worker was within 25-feet of any of the industrial vehicles operating within the facility.

Are the additional controls better? It is hard to measure events that don't happen; however, the number of vehicle strikes declined immediately. Is it perfect? We will see. Chances are this control is much more reliable and effective then either the 10 foot rule or visual observation. Both of these prior actions focused on prevention. The more robust controls focus on recovery. Workers will still need to pay attention because the hazard has not gone away. This example, for this organization, seems to balance prevention and recovery a little more evenly, and from this action the organization will most likely see much better outcomes.

All of these more robust controls were a direct result of the three-question risk assessment technique. This technique is great in reinforcing good systems and even better in identification of insufficient or non-present safeguards. It's clean, quick, cheap, and supports the experts on the hazards and the control work for your organization. Most importantly, it is a different way to assess the work and manage controls. Seeing the question differently allows for different answers to emerge.

Are workers the right source for this type of important operational hazard and safeguard assessment? My answer would be fairly fundamental. Where else would you go to get this type of credible and accurate information? More importantly, where else would you go to ask the sufficiency question? Who else would be able to actually answer if the current controls really do give the worker the ability to fail safely?

I would challenge your organization to dig deeper at the worker level to determine where the actual high-risk operations really exist in your facility and work environments. Knowing what the most risky work is, knowing how we currently manage that risky work, and finally knowing if the controls and safeguard are robust, is important knowledge for any organization to constantly foster and pulse from within the confines of the workplace itself.

Chapter 10

The space where workers manage risk

Hazards are fluid, always in motion. Hazards are never fixed, frozen or stable. The things that hurt people are almost never the things we identify, because hazards appear and disappear as the worker does work. In reality the hazards that workers "manage-around" are constantly changing based upon the context and conditions of the work itself. Workers must adapt to the changing work conditions they face while executing work.

One of the major challenges our industry faces is the idea that risk is somehow fixed in time and space, permanent. We have designed processes that actively try to identify risk as a part of event prevention. We identify all hazards in the work planning stage and then mitigate the hazards in order to keep the worker away from the harm. Most of our tools to manage hazards assume the idea that hazards are fixed. Ask any of your workers and they will tell you the idea that once you identify the hazard you are done managing that hazard is not realistic. Hazard identification is a constant process, not a pre-job task.

Not all risk is the same just as not all work is the same for the worker; workers manage hazards as they appear in real time and as work happens. This process is fluid and almost refuses to be proceduralized or formalized. Hazards require both awareness and adaptability, and workers must tailor their work activities for these constantly moving targets – or maybe the workers are the targets for the hazards – either way the idea that work is ever "as planned" is a brief and fleeting notion to be sure.

I started to look at how workers successfully manage hazard--you know what I mean--observing successful work, work where there is no serious event. I started to notice that the workers don't really get the opportunity to manage the hazard like our planning assumes. What the workers seem to manage is the space between the actual work and the multiple hazards. This seemed interesting to me because it is counter to what we assume will happen when we plan work.

Remember how we discussed risk earlier in this chapter--risk is the degree to which the worker is faced with operational uncertainty. One way to reduce uncertainty, or increase certainty if you would prefer a more appreciative concept, is to manage the margin between the work and the hazard. The more space I have between the bad outcome and me, the less likely I am to have the bad outcome.

Workers tend to keep space between themselves and the hazards that can cause some type of harm. This "space keeping" in reality is an example of risk competency. Workers don't manage the work because successful work requires adaption, expertise, and creativity and workers don't manage the hazards because the hazards are constantly moving in and out of the work environment. That leaves the "space" between the actual work and the actual hazards as the discretionary place where workers create safety. That space could also be called the capacity to do safe work; in fact, let's call this "space capacity" to see how it fits.

What I have just discussed is a different way to think about how workers manage risk or more precisely the capacity between the work being done and the hazards to which the workers are being exposed, but there is more – because there is always more. The capacity doesn't seem to be just one type of capacity. In fact, depending on what the worker is doing, or better yet thinking while the worker is doing work, this safety capacity seems to take on different characteristics of protection.

There is a clear way workers manage this safety capacity as they prepare to do work. There is a much different way the workers manage this safety capacity as they are actually performing or executing the work. These multiple forms of creating safety capacity were interesting to observe in successful work iterations. My bet was these different safety capacities would also be interesting in understanding a failure of any significance.

The Three Types of Capacity

High reliability scholars have understood these different safety capacities between danger and safety for a long while. Reliability tends to match the capacity to operate safely and reliably to the potential disaster or failure that either can happen or has happened. Taking their idea of safety capacity types and applying their thinking not to the matching capacity to the unwanted outcome, but in fact applying their idea of matching capacity to the execution of work.

Now this is getting interesting.

There seems to be three different places where risk requires capacity for workers doing work. These three places are naturally present in the way work is performed. However, the three types of capacity are different, just as these three work stages are different. It might be easier if we introduce the three places that require capacity when doing work.

The worker must manage the capacity to do safe work at every stage of the work execution model. Workers must:

1. Plan to be safe. We call this prevention.
2. Perform or execute work safely.
3. Be able to recover if something fails. We will call this recovery.

Each of these three work capacity areas represents a specialized form of safety thinking that must be done by the worker in order for the worker to safely and productively accomplish work. Your organization's workers manage these three capacities all the time; you manage these three capacities all the time. This is how work is safely completed. Every worker to some extent manages prevention, execution, and recovery – the question is does the worker manage these three capacities with equal importance and attention.

The three forms of risk capacity are all different in importance and complexities, yet each of the three capacities must be present when performing high-consequence work. The three capacities are not equal, at least not in the mind of the person doing the work. In fact depending on what the worker is doing or thinking, the worker may be spending time thinking about execution only. At other times the worker may be completely involved in prevention and planning the work. For high-risk work with uncertain outcomes, it is a good bet the worker is spending time thinking of escape routes and protections needed if the work fails.

Yet all three of these types of risk capacity are equally important. Spending time on one type of capacity and not spending time on the other types of capacity is like giving away two-thirds of your protection. You made not need all three types of capacity, but if you do need this capacity you want the capacity to be there waiting for you. I know this might sound complicated or unrealistic, but I would invite you to really think about how you perform safe work, I will absolutely guarantee that you will see these three types of protective capacity present is stable and safe work.

It is like when you invest in retirement fund. You don't put all your money in risky stocks. Part of your money is in secure and stable money-market funds. Some of your money is in cash. You spread the capacity (your money) not the risk – the risk is there whether your 17 dollars is invested or not.

This observation that we don't manage risk, we manage the capacity we expose to risk (or really hazard) really changed the way I started looking at fatalities and serious events. Perhaps what we have is not bad workers making bad decisions, nor do we have bad work (the hazards again) that is trying to kill our workers. Perhaps what we have here is an absence of capacity at one of the three safety capacities.

If we have all our capacity on prevention, we would have little capacity for work execution and even smaller capacity for recovery. I would guess the transverse of this would be true. If we put all our eggs in the recovery basket, we would have little prevention capacity in order to safely do our work.

It strikes me that you could represent this idea using a picture of a scale – you know, like the scales of justice. A scale would show the need for a balanced approach to managing capacity at the organizational level with one huge exception – a scale normally is used to show a balance

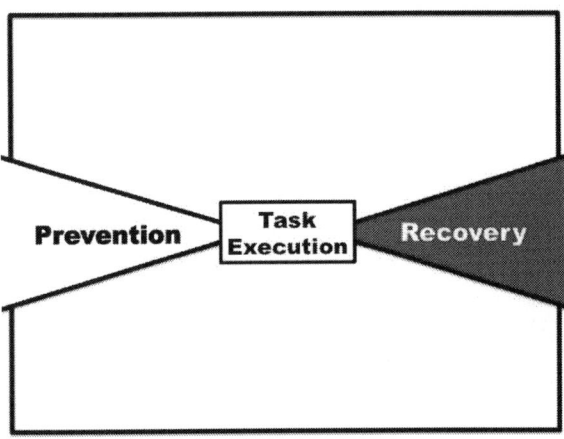

between two values and we have three capacities. That has led me to represent this idea in my organization as a bow tie model. You can show a balance in the sides of the bows and the middle seems perfect to represent the capacity to execute work.

However, our discussion thus far has worked hard to make the claim we give greater preference to one of these three types of safety capacity over the other two. In fact, I will stake this books reputation that we think prevention capacity is much more important then work execution or recovery capacity. Our scale is out of balance. Our bow tie is fat on one side and small on the other. We are amazing at prevention and almost capacity-less on the recovery side.

This imbalance between prevention capacity and recovery capacity is great when work is successful. It actually is even a bit "self-fulfilling" in that it looks like all the effort and resource we put into prevention actually paid off handsomely in the complete presence of production success and the complete absence of fatalities and serious events.

Think about our illustration case studies. Think about our poor movie crew, and think about serious events in your organization, and it is quite probable that you will find an imbalance between the capacities. The film crew seemed to put all their effort on shooting the scene on the railroad trestle. So much capacity was used to create this gorilla cinema that almost no capacity was spent on prevention capacity or recovery capacity. You could make a similar case on almost any case study you chose to use--the balance between prevention, execution, and recovery seems almost vital to doing stable and safe work.

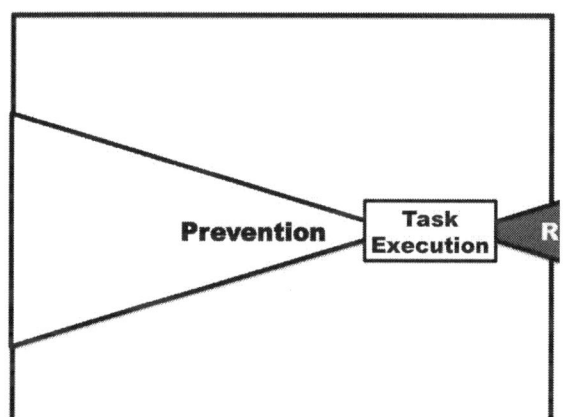

The best proof is when I observe high-performing teams, teams that are productive, efficient, stable, and safe. These high performance teams seem to naturally understand that you manage capacity in a balanced approach. When a tugboat goes in to the Harbor of Long Beach to provide ship assist, the crew manages prevention by ensuring the vessel is ready and prepped for the job. The crew of course executes the work in real time always detecting and correcting for the changing work and work hazards. Most impressively, a tugboat captain is always planning for where the tugboat will end up as the tugboat is executing work. Like a good billiards player, the tugboat captain must always position the tugboat for the next task to be preformed while performing the current task.

The tugboat crew may not realize they have a balanced approach to managing safety capacity at all three levels of work resilience, but they absolutely do this work every time they perform their mission. This idea of multiple different safety capacities is not new; I am just not sure we have named it or thought of it as a way to address events. This became clear to me in the midst of all these fatalities peer assist visits I was doing. I was in a role where I was allowed to look at really safe and stable operations that had just suffered a catastrophic failure, and I began to notice a strong prevention culture but not a strong recovery culture.

I see the same examples with linemen working a storm response. These linemen manage capacity at all three levels because the work is so uncertain and unpredictable. Lineman always manage prevention before the go to the pole, they execute with capacity because of the hazards with which these guys and gals interface, and during a storm recovery-capacity is the entire reason they can do this high-risk work. Everything the lineman does during a storm is built upon the foundation that this task could screw up at any moment and when this task does screw up he or she wants to be where they have the most protection and the least exposure.

These and other examples of successful high-risk work is how a balanced approach, equal capacity at all three levels of work, shows itself in practice. I challenge you to not find examples of good teams managing multiple capacities when you look at your organization. We know this practice exists informally among seasoned experts who have done dangerous work for years. Could it be time to perhaps take a more formal approach to purposely creating this balance?

STOP Work

We have reached the end of the usefulness of STOP work. I am not sure

STOP work was ever effective, but it felt like the right thing to do, so we did the crap out of it. I would guess the data collected by observers when a worker was observed stopping a job before it failed and injured or killed someone is totally anecdotal and not scientific.

You know, of course, if your organization counts on STOP work as a prevention strategy that means that every other prevention strategy that your organization uses failed. Really using STOP work as an explanation for why a bad outcome happened is perhaps the weakest (and dumbest) excuse and organization could have for why a bad outcome happened.

I am not saying STOP work is wrong. It just strikes me as painfully simple and incredibly retrospective. We tend to use the STOP work criteria as a finding in an investigation and never a tool to execute safe and stable work. Asking workers to STOP a job before the job has bad consequences feels like management is creating a "back door" that removes all responsibility from the organization. When something fails, it would be easy to simply say, "The worker should have stopped the job." Management is relieved of the responsibility for creating work that can be done safely and the accountability for the accident goes directly on the workers who, not only failed to stop work, but also had an accident.

Throughout this entire discussion so far, we are building a case that prevention is not sufficient to manage serious outcome events like fatalities and serious events. We know this because organizations with amazingly formal and culturally rich prevention process still kill workers and have serious failures. These more catastrophic events seem to defy our prevention models at every turn. Try as we might, the events that cause fatalities happened in spite of all our efforts to prevent them. Prevention does not seem to be the most correct choice for an attempt to curtail serious accidents.

That said, at its best STOP work is a prevention tool. STOP work is not, nor ever has been nor ever will be a control. STOP work is supposed to happen before the undesired outcome or consequence happens. That is the entire point of STOP work. STOP work is supposed to stop the job before someone gets hurt. IF this use of STOP work is accurate, then STOP work authority is a prevention tool and not a control or a safeguard.

I really hate to admit this, but I had never thought about this nuanced view of STOP work until I was doing a learning team with a group of workers who had had a fatality. This crusty old mechanic looked me straight in the eye and told me, "STOP work is at best a prevention tool and not a good prevention tool at that. " That stalled the conversation for a bit. I nodded and acted like I knew that, but inside I was processing this information as fast as I could. That crusty mechanic is exactly right, he could not have been more right, and he should make you think a bit as well.

So what does all this mean? It seems a bit terrible to say this process is bad and offer no alternative. Let's establish that STOP work is ineffective and abusive towards your organization's workers. STOP work doesn't really prevent events, because if the workers knew the action they were about to take would cause harm to them or their coworkers, they would stop work whether they had the authority or not. STOP work is used after a consequence to ask workers why they failed to use their STOP work authority. STOP work dumbs down the complex and context rich conditions that create environments that foster accidents. They make an accident look like it would be as simple to avoid as saying the word, "stop."

Instead of asking worker to psychically identify the precise uncertainty that will lead to an accident, why not push the effort towards ensuring the right controls and safeguards are in place for the work to fail safely *before* the work starts. Perhaps STOP work should be called START work. Why not empower the workers to have the power to not start a job if the appropriate safeguards are not present or engaged.

Lets move from placing our operational emphasis on STOP work to placing renewed operational emphasis on START work. Every worker has the power to not start the task until the workers are assured the proper safeguards are in place to allow for the task to fail without consequence to the workers or the facility. Can we fail safely and gracefully? Does this feel more in line with the shift in thinking that we have been discussing throughout this book?

This START work idea removes the pressure on prevention and places this pressure on the recovery capacity that must exist to successfully execute potentially high-consequence work. The jury is out if this idea will work. START work still has a slight tinge of "weaponization" if used retrospectively after a failure, to be sure, but does directly address the shift toward creating a balance between prevention and recovery efforts while doing work.

All of these efforts indicate a new way of thinking, not necessarily a new way of doing work. Highly effective work groups have been managing safeguards for years. If you watch the successful execution of high-risk work by really experienced and skilled workers, you will see these workers address recoverability before these workers ever touch a tool.

Chapter 11

What we know now – as a result of this discussion

The second half of our discussion has been presented to introduce different ways to think about and counter the forces that are stopping our organizations from not making progress on the deliberate reduction of fatalities and serious events that we so deeply desire. Our current thinking and processes about the ways we reduce fatalities and serious incidents has not created the outcomes that we want, in fact the data shows that the rate of fatality reduction is not reducing at the same rate as the reduction of loss time events and other industrial safety indicators.

We know that we cannot solely rely on prevention as planned, just as a worker cannot solely rely on procedures and planning to completely represent the work that is actually being performed. We know that prevention alone is not sufficient to interrupt an event with fatal consequences; we also know that we would be foolish to only count on controls and safeguards.

- *Our current safety systems are not sufficient to prevent high-consequence outcome events.*

Asking why our current systems and processes to manage industrial safety do not seem to translate directly in to the management of more serious outcome events is really the first step of the change that is happening around this problem. We recognized the current failing by the mere asking of this new question. "How come we are good at ankle sprain reduction and bad and not killing people?" Doing our work differently is a challenge for so many reasons and thinking about our work differently is, perhaps, the most challenging struggle of them all.

It is important to know that we cannot abandon, nor would we ever want to abandon, planning and prevention as a strategy to preform high-risks tasks-reliably. However, we know prevention activities alone are not sufficient. Our organization must also provide the consequence management against catastrophic failure of all types. We cannot rely on prevention and attention, alone to create safe and stable work.

Knowing our thinking is changing because our questions are changing is a positive indicator and not a point of frustration. We must happily ask new questions and push for new answers. It is only through the new thinking that the different tactics and activities will be discovered. Reinforcing the same preventions does not make our processes more robust.

- *Recovery will always meet three tests. The controls must be present. The controls must be robust. Finally, we have to be to imagine, not the potential event or uncertainty, but the failure-consequences (potential outcome) that will be present.*

We cannot predict the almost infinite ways that a process may fail; we can predict what the outcome of the failure could be. I can't tell you why this process will explode, but I can tell you what would happen when an explosion of any origin happens. We can't predict what will fail, but we can predict what a failure will do to our systems and processes.

The introduction of three correlated reasons people die in fatal accidents helped to illustrate and introduce the shift from the old "failure to prevent" the accident thinking to a much different "failure to control" mode of thinking and responding to events. I told you that of all the types of fatal accidents in my 2016 data set, I was able to bucket the events into three categories of recovery failure:

1. Absence of controls.
2. Insufficient controls.
3. Events beyond safety basis, events beyond risk imagination.

In our time together, we started, only really just started, to discuss the actions that happen at the front of the accident (prevention) and the actions that happen after the consequence at the back of an accident (safeguards and controls) as a way to understand and create new mental models. The new models are less linear, for serious events and fatalities that allow our organizations other potential intervention points, beyond planning and prevention, where we can have reduction impact on serious consequence events.

These categories are not prediction tools. Although these three tools may have direct positive impact for safeguard and control verification, these categories were much more a product of event analysis post event. Knowing why the controls were not present is extremely helpful in knowing how to respond to an event. Knowing the controls were not present, not robust, or the event was beyond imagination becomes an important part of the path forward post consequence. These categories help create corrective actions and event resolutions.

- *High-risk work with robust and mature controls is less risky then low-risk work without controls.*

The discussion introduced some more current thinking on risk assessment and controls. The most interesting realization for me was in identifying that those high-risk activities that have mature and robust controls are really not as risky as low-risk activities that do not have mature or robust controls. Sometime these low-risk activities have no controls at all. That is why we are so good at doing super risky work, and we kill workers when we chop down trees or move pieces of industrial equipment across the plant floor.

To be honest, this learning for me is why I became so interested in serious events and fatalities. Knowing high-risk work with mature controls is safer then other types of work without controls makes complete sense. The remarkable thing about this statement is that I had never heard this statement said before in any workplace. Our attention towards making high-risk work mature and stable is why this high-risk work has mature and robust controls.

We understand the importance of controls; we are simply fixated on the idea that prevention is better. So much so that we often rely entirely on planning and prevention to make work stable.

- *Asking workers to help assess the presence and effectiveness of controls and recovery is engaging, quick, enlightening, and accurate.*

We introduced this new risk and control thinking as a way to have hazard and risk conversations with the workers who do the work. The realization the workers are not the risk-control problem, but in fact workers are the risk-control solution gives our organizations an effective way to assess both the presence of safeguards and the quality of safeguards simultaneously. By simply asking workers three questions:

1. What work, when our systems and processes fail, will kill you?
2. What keeps you from getting killed when the process fails?
3. Are those controls and safeguards enough?

We know there is not a better population to engage in this conversation. The benefit from simply asking workers to identify the risk-consequences (potential outcomes) in actual work is remarkable in and of itself. Knowing if the controls and safeguards are sufficient is probably the difference between a catastrophic failure and a safe failure.

Most importantly, this engagement of the workers who do the work empowers workers to have the confidence to manage and identify controls. The workers, after all, cannot manage the absence or presence of an accident or unexpected event. Remember, accidents and unexpected events are accidents and are, by definition, unexpected.

- *There is a balanced approach to managing both prevention, work execution, and safeguards and control. Without a balanced approach, the work has a higher potential to fail catastrophically.*

Finally, we built a drawing to show the balance that is needed between our efforts before the accident, the prevention work we do in our organizations, and the recovery safeguards that we put in our work processes after the consequence has happened. Both prevention and recovery are surrounding the third type of protection our organizations rely upon which is the protections the worker always manage while executing the work.

By using this bow-tie drawing and tools designed to assess both prevention and control, we are able to show a possible imbalance between prevention and control. Anytime one bow of the bow tie is larger then the other bow of the bow tie we know we have a work process system that is out of balance. We manage both bows of the bow tie for different reasons just as we manage both sides of work execution for different potential outcomes.

Knowing you manage your organization's capacity to fail safely by managing the balance between prevention, execution, and recovery multiplies your potential management opportunities. Your organization no longer simply manages planning and prevention. You now manage a balanced approach to failing safely by management of controls in parallel with prevention. You don't really get to manage execution safety as an organization; the worker in the adaptive work environment always manages work execution.

Chapter 13

The Mission: Stop Killing Workers

We have worked our way through an important discussion about fatalities and serious events. We have traveled in this discussion through the unique characteristics fatality and serious events present to all organizations that desire reliable, safe, and stable operations. We know catastrophic outcomes are outliers, and we know this because mostly our organizations do not suffer catastrophic losses or outcomes; these events are not normal outcomes. We have discussed the bias that are clearly built-in to human thinking that lead us to believe the cause of a fatality or serious event was the organization's failure to prevent the catastrophic or serious event. We also know your organization, no matter how big or great you are, will never be completely successful at preventing all events.

This journey is in most ways about shifting the way we view the problem of fatalities and serious events. The best way to change our solution to this difficult and important problem is to change our problem statement. We know in order to get different answers we must ask different questions. If we always see these accidents as an indicator of a major deficiency in our safety program, we will always be tempted to fix this problem by reinforcing our existing thoughts and beliefs, repairing the safety program deficiencies, that make up the fundamental definition of our existing prevention program. In short, we will try to do the same things – better.

If we see fatalities and serious events not as a failure of prevention, but as a failure of a balance approach of prevention, work execution, and controls, we then are duty-bound to dramatically improve our safeguards, creating systems and processes that have the capacity to fail safer, or in the best case scenario fail safely. We, at best, manage our system's ability to offer workers more room for recovery. We know non-recoverable outcomes kill workers; the data on workplace fatalities are clear and illuminating.

The lessons are there; we must but learn these lessons well. Let me give you an example that started my thinking about this specific problem many years ago. I knew organizations got into trouble when those organizations did not give workers the space to think and respond. I also was starting to think the workers were not thinking and responding because we did not give them space to allow the thinking and responding to happen. Not dying or getting hurt is not from a lack of will or skill on the part of the workers. It seemed that workers are often surprised by outcomes and left with little time to think and respond their way out of the trouble. Workers need some capacity to respond effectively.

Another orbit around the planet

Creating the capacity to respond effectively is not a problem of behavior; capacity is a problem that is always owned by the process. For example, if the space program is going to land the space shuttle, remember this is many years ago and so this example is a bit dated and inaccurate but should suffice to illustrate the power of having more capacity built in to our systems. The mission leadership has designed into their shuttle landing process the capacity to solve problems before these problems become catastrophic. Mission control builds in capacity before the unexpected event happens.

If for some reason there is a problem with landing the space shuttle, the space shuttle always maintains the capacity to orbit the earth one more time. One more orbit around the earth buys time; time allows for thinking and problem-solving, problem solving solves problems. This built in capacity is there if it is needed; in fact, this capacity is always there if it is needed. If things get really bad, something unexpected happens, there is always the option to take one more orbit around the earth. The orbit gives the designers, workers, mission leadership, even the astronauts some margin between the problem and the outcome.

Commercial planes generally put enough fuel on board the plane to fly to the airport that is on the flight plan, the original destination, and enough extra fuel to get to several airports near by the original destination airport. This extra fuel is on board for the sake of capacity, if something unexpected happens, there is space and time to make decisions and solve problems. As a person who flies often, that capacity to recover is important to me. Failure happens in planning and logistics for airlines and at airports because of weather and schedules all the time. If commercial aviation always failed non-recoverably fewer people would fly on planes and there would be fewer planes to carry the brave daredevils that did fly.

How different our case study on the death by train strike of Sarah Jones would have been if there were to have been space to stand when the unexpected combination of a speeding freight train and a tangled bedframe created an environment where an unsurprising consequence tragically happened to a crew of surprised workers. The belief that this event should never have happened, which by the way is completely true, is not the same belief as realizing the crew had no other options once the event had started, there was no additional time or space in which this crew could safely solve the problem.

These examples are direct operational responses to the same type of problems that you have, unexpected events – outliers – anomalies. We know, because we have had a rather elaborate discussion on this topic, we cannot predict unexpected events. We cannot manage uncertainty by managing probability and prevention. We must manage uncertainty by management of capacity, one more orbit around the earth or enough fuel to get to the next airport.

Knowing capacity is the key, a balanced approach to prevention, work execution, and recovery, we are now duty bound to ensure this capacity is in our work processes. For the organizations that have had failure, this is a fairly easy conversation to have, for organizations that have not had some type of serous event or fatality event, this conversation is a bit more difficult. In a way you are almost forced to talk about capacity before the identified need for increased capacity makes sense beyond prevention. Stay strong and carry on. Once your organization's leadership begins to understand the idea that they must see safety differently, the conversation becomes much easier. Eventually, your leadership will be challenging you to think deeper, to think newer, to think beyond prevention.

We don't have a failure to prevent a serious or fatal accident in our organizational processes and systems. We have a failure in our *belief* that we can predict all fatalities if we would simply try harder. We have in imbalance between our ability to plan and prevent the event, the safety that workers provide while executing the actual work, and the ability to safely fail when the unexpected and unplanned happens.

Postlude

You have two choices: Getting Better or Getting Even

The Power of Restoration

When a fatality or catastrophic failure happens **how** the organization responds to this event is vital to the organization's ability to recover from this event. Your organization now knows you are not going to get to zero – you will not be that "event free" operation that you have always hoped you would become. A really bad outcome has happened and the past is horrific, the future is even more frightening. You can't change history. You must change the future. Yet, you are in the present time and you are desperately trying to understand both what happened and what should happen next.

The work must continue in spite of the catastrophic event. The work must continue *because* of this catastrophic failure. Somehow, your organization must figure out the path forward through a very emotional and complicated. Decisions must be made. How should we respond to this horrible event? How should we proceed in the face of the ultimate uncertainties? What is the right response? What is the right reaction?

Your organization's response to this horrible event is a deliberate management strategy. There are many victims of a catastrophic failure. The workers that have been injured are the first victims. Others in the organization are also victims of this failure. You soon realize that

> **Organizations must choose between Getting Better or Getting Even.**

although you are also among the second-level victims of the catastrophic outcome, you will be called to lead the organizational response to this horrible failure. You have to determine what tomorrow will look like for the organization and its members, how tomorrow will sound, and how you begin to restore your organization's ability to do high-risk work even better then you did before the failure happened. You will be called upon to make your organization better.

Let me state this sentence again, no matter how obvious the following sentence may sound: When a fatality or catastrophic failure happens **how** the organization responds to this event is vital to the organization's ability to recover from this event. Your reaction and responses, both immediately and for the long-term, will set the course and direction for how your organization restores its ability to do high-risk work even better then you did this work before the event.

Restore. The key word is restore. Not repair, not react, not respond – but restore your organization's ability to do high-risk work safer then your organization has done this work in the past. Our task is to restore a group of good workers who are hurting to a place where these workers will once again become brave and effective. You want to restore your organization to a place beyond its former state. The old view, the old way, of responding to an even does not help you restore your organization back to a higher ground. You must get to higher ground.

You will not be able to blame your way back to normal. Simply determining that the accident happened because a person failed to prevent the accident from happening is not going to cut it. Your organization will never be the same, it cannot be the same, and your organization has been hurt and will need to mend. You don't want your organization to be the same – the same organization gave you the failure that changed everything. You will, and must, begin the process of restoration.

A restorative response to the aftermath of a catastrophic failure asks three very important questions:

- Who is hurt?
- What do these people need in order to restore their ability to do this work better then it was done in the past?
- Whose obligation is it to meet those needs? Who is going to take the effort and accountability (used purposefully in the forward sense of this word) to restore your organization's operations to a place beyond the place where operations once where.

As Dekker says in the introduction to this book, "such approach is not only more 'just,' but also more inclusive." Many people are affected by an incident: not just the first victims (e.g. the worker) but also colleagues, supervisors, bystanders, the organization, the surrounding community—they too may somehow have been affected by what happened. If, as fellow human beings, and especially as leaders, we realize that hurt creates needs, and that needs create obligations, and obligations create accountability, then we can take our responses and conversations in a very different direction.

Catastrophic events create the need for a new and better awareness of the holistic needs that your organization and its membership has and then the act of addressing those needs become vital to the organization's ability to restore its operations *beyond* where the operations began its work before the failure had consequence. Restoring the confidence and capacity to do work safer, better, and more effectively – while helping give the people in your work community what they need to go on.

Restoration is an important job. Restoration doesn't spend time explaining the hurt in your organization; restoration spends time learning and understanding what the organization needs. Time is spent not in assigning blame, but in learning what the people need.

The organization must provide for the needs that have been identified by the catastrophic failure. This event, this horrible thing that has happened, is information rich and must be understood – beyond our traditional investigation and blame-laying rituals that have not served us well in the past.

Organizations must shift thinking from operational recovery to holistic restoration. Don't look for a single item, a root cause if you will, instead seek the things you can do to make your organization's workforce feel whole again. Search for ways to learn and understand what workers know about ways to control consequence after an operational failure has taken place. What can we do to help the workforce get better, get more stable, become whole again?

Again, learning about operations from the people who do the operations, allows your organization to better understand what the workers will need in order to restore work to a more stable and recoverable place. Knowing these workers both need to recover and move forward – at the same time – gives the organization a chance to both stabilize and improve.

Restorative approaches succeed by systematically considering these needs, and working out collaboratively whose obligation it is to meet them. Restorative responses to fatalities or significant failures seek to honor the horrible event by getting operationally wiser. When fatalities do happen, and an organization is reeling from the shock and grief, the last thing we should do is to revert to blame. Getting even is not getting better. Our journey is to not get back to business as usual. Our journey is to restore our operations to better then these operations were before the event.

ABOUT THE AUTHOR

Todd Conklin spent 25 years at Los Alamos National Laboratory as a Senior Advisor for Organizational and Safety Culture. Los Alamos National Laboratory is one of the world's foremost research and development laboratories; Dr. Conklin has been working on the Human Performance program for the last 15 years of his 25-year career. It is in this fortunate position where he enjoys the best of both the academic world and the world of safety in practice. Conklin holds a Ph.D. in organizational behavior from the University of New Mexico. He speaks all over the world to executives, groups and work teams who are interested in better understanding the relationship between the workers in the field and the organization's systems, processes, and programs. He has brought these systems to major corporations around the world. Conklin practices these ideas not only in his own workplace, but also in the event investigations at other workplaces around the world. Conklin's best selling books, Pre-Accident Investigations: An introduction to Organizational Safety, and Pre-Accident Investigations: Better Questions are among the most read books on new safety. Conklin lives in Santa Fe, New Mexico and thinks that New Safety and Human Performance is the most meaningful work he has ever had the opportunity to live and teach.

Printed in Great Britain
by Amazon